WATER AND SUSTAINABILITY

Edited by **Prathna Thanjavur Chandrasekaran**

Water and Sustainability
http://dx.doi.org/10.5772/intechopen.73414
Edited by Prathna Thanjavur Chandrasekaran

Assistant to the Editor(s): Ali Ajaz

Contributors

Ketlhatlhogile Mosepele, Wame Hambira, Goemeone Mogomotsi, Patricia Mogomotsi, Oliver Moses, Masego Dhliwayo, Anastacia Makati, Bernice Setomba, Josh Sperling, Will Sarni, Jean-Philippe Terreaux, Jean-Marie Lescot, Elena Kim, Gurdeep Singh, Mansoor Leh, Dan Li, Mngereza Mzee Miraj, Prathna Thanjavur Chandrasekaran

First published in London, United Kingdom, 2019 by IntechOpen
IntechOpen is the global imprint of INTECHOPEN LIMITED, registered in England and Wales, registration number: 11086078, The Shard, 25th floor, 32 London Bridge Street
London, SE19SG – United Kingdom
Printed in Croatia

British Library Cataloguing-in-Publication Data
A catalogue record for this book is available from the British Library

Additional hard copies can be obtained from orders@intechopen.com

Water and Sustainability, Edited by Prathna Thanjavur Chandrasekaran
p. cm.
Print ISBN 978-1-83880-231-8
Online ISBN 978-1-83880-232-5
eBook (PDF) ISBN 978-1-83880-635-4

For EU product safety concerns:
IN TECH d.o.o., Prolaz Marije Krucifikse Kozulić 3, 51000 Rijeka, Croatia,
info@intechopen.com or visit our website at intechopen.com.

Meet the editor

With a PhD in Nanotechnology, Prathna Thanjavur Chandrasekaran has been working in the field of drinking water treatment for the past 5+ years and has more than 28 publications in peer-reviewed international journals and 1300+ citations to her credit. She has worked with academic institutions of repute such as the Indian Institute of Science (Bangalore) and IHE Delft Institute for Water Education, the Netherlands (formerly UNESCO-IHE Institute for Water Education). Her research at these institutions primarily focused on optimizing adsorbent nanomaterials for the development of decentralized drinking water filters. She recently joined the Department of Irrigation and Flood Control as a Chief Minister's Fellow with the Delhi Government (India) and works on projects related to the revival of water bodies, water management, and reuse.

Contents

Preface IX

Section 1 Water Dynamics and its Impact 1

Chapter 1 **Introductory Chapter: Water Sustainability in a Dynamic World 3**
Prathna Thanjavur Chandrasekaran

Chapter 2 **Sustainable and Resilient Water and Energy Futures: From New Ethics and Choices to Urban Nexus Strategies 7**
Josh Sperling and Will Sarni

Chapter 3 **A Call to Cities: Run Out of Water or Create Resilience and Abundance? 25**
Will Sarni and Josh Sperling

Chapter 4 **Water, Ecosystem Dynamics and Human Livelihoods In the Okavango River Basin (ORB): Competing Needs or Balanced Use? A Review 41**
Ketlhatlogile Mosepele, Wame L. Hambira, Goemeone E.J. Mogomotsi, Patricia K. Mogomotsi, Oliver Moses, Masego Dhliwayo, Anastacia Makati and Bernice Setomba

Chapter 5 **Sustainability of Irrigation in Uzbekistan: Implications for Women Farmers 65**
Elena Kim

Chapter 6 **Comparison of Water Resources Community Self-Management Mode between China and Tanzania 79**
Dan Li and Mngereza Mzee Miraj

Section 2 Water Dynamics and Economics 93

Chapter 7 **Economic Instruments to Combat Eutrophication: A Survey 95**
Jean-Philippe Terreaux and Jean-Marie Lescot

Chapter 8 **Setting Up a Computer Simulation Model in an Arkansas Watershed for the MRBI Program 113**
Gurdeep Singh and Mansoor Leh

Preface

Floods, droughts, and famines brought about by the vagaries of nature cause damage to the livelihoods of people. The need to adopt sustainable practices toward water use, reuse, and management is relevant in the current scenario and the scope of this book deals precisely with sustainable practices in water management. The purpose of this book is to provide the reader with abundant and relevant information on all aspects related to water sustainability: water reuse, dynamics of transboundary waterways, economic tools to analyze sustainability, water-energy-food nexus, computer simulation models to study watershed models, and so forth.

The book is divided into two parts. The first part of the book discusses aspects related to the dynamics and sustainability models adopted along water bodies. The chapters by Josh and Will, for example, touch on an important aspect of water scarcity that is affecting major cities in the world. The chapters elaborate the need for sustained choices to combat the risk of droughts and floods in cities. Cape Town is the first major city in the world to encounter water shortage after a three-year drought and many other major cities are not far behind. Therefore, the need for sustained choices in water management is all the more critical. The chapter on the Okavango river basin whose waters are shared by Namibia, Botswana, and Angola discusses the challenges of water sharing and transboundary waterways along the river basin. Livelihoods and dynamics along the stretch of the river basin ecosystem are described beautifully in the chapter by Ketlhatlhogile. Gender aspects of the implications of irrigation are discussed in detail in the chapter by Elena. The chapter briefs on the implications sustainable irrigation has on women farmers in Uzbekistan. Management of water resources is far more critical in rural arid and semiarid areas because it can directly impact the livelihood of the rural community. The last chapter in this section by Li and Miraj compares the water resources community self-management mode based on case studies in rural arid areas of China and Tanzania. The authors in this chapter reveal that the self-management mode was primarily driven by village governments in China, while it was largely driven by nongovernment organizations in Tanzania.

The second part of the book focuses more on simulation models and survey studies to determine economic instruments to counter nutrient enrichment in water bodies. Nutrient enrichment or eutrophication of water bodies negatively impacts the aquatic ecosystem by disrupting all levels of the food chain. Eutrophication is an enormous issue that ails most of our water bodies and has a potential to impact goods and services in the long run. Eutrophication is caused mainly by surface runoff of fertilizer-laden water from agricultural fields and release of untreated water from industries and housing colonies into the water body. The chapter by Lescot provides an interesting survey insight into the economic aspects of eutrophication in the context of France and Europe in general. The aim of the survey chapter is intended to help in public decision-making in reducing eutrophication. Field studies with regard to studying best management practices to maintain water quality can be laborious and time consuming. Simulation models can go a long way in providing accurate results and be less labor intensive. In the same vein, the chapter by Singh and Leh calibrates and validates a simulation model in an Arkansas watershed. The chapter provides a detailed insight into various steps for data preparation before calibration and validation of the Soil and Water Assessment Tool model, which is the widely used model to assess the impact of various best management practices.

Although this book may not provide readers with comprehensive information on all aspects related to water sustainability, it will provide constructive data and content on the current trends and advancements in sustainable practices related to water. The book is intended to further motivate readers and scientists alike to look further and make concerted efforts toward promoting better and effective water management.

Prathna Thanjavur Chandrasekaran
Department of Irrigation and Flood Control
Govt. of National Capital Territory of Delhi
Delhi, India

Section 1

Water Dynamics and its Impact

Chapter 1

Introductory Chapter: Water Sustainability in a Dynamic World

Prathna Thanjavur Chandrasekaran

Additional information is available at the end of the chapter

http://dx.doi.org/10.5772/intechopen.85150

1. Introduction

Water is the largest limited natural resource which is vital for survival of all living beings. Floods, droughts, and famines brought about by climate changes have been noted to occur with more frequency in the recent years. Therefore, the need for the adoption of sustainable methods toward water use and management is critical in the present-day scenario. In addition, there is also an urgent need to develop policies and make smart investment decisions to promote water sustainability in the light of climate change.

2. Water footprint

Increase in standard of living around the world has also brought with it an increase in the demand for water-intensive goods further impacting the already limited fresh water resources [1]. Water footprint is the term used to calculate the amount of water pertaining to the production of a commodity. The water footprint can either be classified as green, blue, or grey depending on the source of water used for producing a product. While green refers to the amount of rainwater, blue refers to the amount of groundwater used to produce a product. On the other hand, grey water footprint refers to the amount of fresh water that will be used to dilute an aqueous solution consisting of pollutants to bring it down to the desirable level. It is critical that countries take steps to keep the water footprint at the lowest and promote sustainable management of water resources as the implications of irresponsible water usage can be global in nature [2].

3. Water-stressed cities

Studies from The Nature Conservancy report indicate that one in four major cities is water stressed, and the demands for water will double in the next three decades. The Water Scarce Cities (WSC) Initiative was initiated by the World Bank to promote water security. This was promoted by increasing awareness among people and nations, facilitating collaborations and dialogs between stakeholders, and by facilitating and providing technical assistance and logistical support in countries [3]. In the present situation of water scarcity, it is critical to rethink related aspects of urban water management in the light of rainwater harvesting, water use, and wastewater treatment since all these sectors need to be handled together [4].

4. Revival and rehabilitation of waterbodies

Water management for a sustainable future also includes rejuvenation and revival of polluted waterbodies. Revival of the waterbody promotes the conservation of the ecology of the surrounding area. Rehabilitation of waterbodies is first brought about by cleaning/removal of the sludge settled, followed by methodical ways to stem the flow of effluents from industries or raw sewage into the waterbody. Recently, there has been an increased interest in exploring eco-friendly approaches in treatment of waterbodies. In the same vein, the potential of floating wetlands in promoting water purification has been widely studied. Floating wetlands often consist of hormonally treated plants with synergistic bacterial colonies in the rhizosphere region which have the ability to adsorb harmful trace contaminants in the waterbody.

5. Promoting clean water and cities

Proper disposal practices with regard to hazardous chemicals (e.g. fertilizers) can go a long way in reducing the load of nutrients reaching the waterbody. Excessive levels of nitrogen and phosphorus in waterbodies (from surface run-off from agricultural fields) can promote eutrophication of waterbodies leading to the destruction of the fresh water ecosystem. Maintaining a low level of water footprint and judicious water reuse strategies can significantly reduce the water consumption and reduce the water stress. Water reuse strategies such as a grey water pool system can be undertaken at a larger scale in communities to promote sustainable water management. Adoption of techniques like rainwater harvesting to recharge the aquifer as well as use it as a source of water can be promoted extensively thereby minimizing run-off of precious rainwater into open drains [5].

6. Conclusion

The 2030 Agenda for Sustainable Development put forth by the United Nations identified 17 Sustainable Development Goals (SDG) which requires urgent attention, implementation, and monitoring by both developing and developed countries. SDG 6 and SDG 14 are focused on ensuring availability of safe drinking water to all and sustainable use of marine and ocean resources, respectively. This explains the urgent need to move toward sustainable water management. Many countries have taken initiatives to adopt sustainable water use and reuse techniques, while others have realized the need for such measures. Protecting the ecosystem for the future generations begins with judicious water management strategies today.

Author details

Prathna Thanjavur Chandrasekaran

Address all correspondence to: prathna.tc@gmail.com

Department of Irrigation and Flood Control, Government of NCT of Delhi, India

† Formerly with Department of Environmental Engineering and Water Technology, UNESCO-IHE, The Netherlands

References

[1] The Importance of Water Sustainability [Internet]. 2015. Available from: https://www.thegef.org/news/importance-water-sustainability [Accessed: February 10, 2019]

[2] Leblanc R. Water Footprint and Its Growing Importance [Internet]. 2017. Available from: https://www.thegef.org/news/importance-water-sustainability [Accessed: February 09, 2019]

[3] World Bank. Water Scarce Cities Initiative [Internet]. Available from: http://pubdocs.worldbank.org/en/588881494274482854/Water-Scarce-Cities-Initiative.pdf [Accessed: February 09, 2019]

[4] Cole LN. Reduce and Reuse: Surprising insights from UC Berkeley Professor Sedlak on What Makes a City More Water Resilient [Internet]. 2017. Available from: http://blogs.worldbank.org/water/reduce-and-reuse-surprising-insights-uc-berkeley-professor-sedlak-what-makes-city-more-water [Accessed: February 09, 2019]

[5] Al-Jayoussi OR. Greywater reuse: Towards sustainable water management. Desalination. 2003;**156**(1-3):181-192

Chapter 2

Sustainable and Resilient Water and Energy Futures: From New Ethics and Choices to Urban Nexus Strategies

Josh Sperling and Will Sarni

Additional information is available at the end of the chapter

http://dx.doi.org/10.5772/intechopen.82825

Abstract

A safe, secure and affordable water future—for life, health, economy—are foundational outcomes from a new form of ethics for water stewardship and energy management. Current business as usual in water and energy systems have not led to sustainable, healthy nor resilient pathways for urban and rural communities alike. Today, an estimated 400 million people live in cities with significant water shortages. This is while 25% of water is currently lost before even used in urban areas (up to 60% in some cities) due to aging infrastructure. In addition, on average, only 10% of wastewater is treated before returning to water bodies in developing countries. By 2040, more than 66% of the world's populations could suffer from severe water shortages; and by 2050, an 80% increase in urban water demand (over current levels) may result in one billion city dwellers and 36% (one in three) of cities expected to face water crises. A crisis is often a catalyst for innovation and this chapter is a call to cities to enable strategic responses—moving away from legacy 'siloed' infrastructures, over-allocated water resources and emerging ethical dilemmas to integrated water- and energy-related urban nexus strategies.

Keywords: ethics, choices, infrastructure, nexus strategies, resilient urban systems

1. Introduction

The world's aquifers are being depleted at rapid rates due to growing populations and unsustainable urbanization practices, including land use sprawl and ever-increasing water and energy resource demands. According to recent analyses, an estimated 400 million people already live in cities with significant water shortages today, and with an 80% increase in urban water demand (over today's levels) projected by 2050, 1 billion city dwellers and 36% of (one in three) cities are expected to face water crises by 2050 [1]. Even earlier, by 2040, more than 66% of the world's populations could suffer from sever water shortage—all while 25% of water is lost before even used in urban areas (up to 60% in some cities), due to poor maintenance or infrastructure. In addition, on average, only 10% of wastewater is treated before returning to water bodies in developing countries [2].

Current lack of reliable infrastructure, high levels of pollution, increasing frequency and intensity of extreme weather events (e.g. droughts, flooding, wildfires), power outages for water and wastewater utility services in and around cities—all these factors will continue to affect reliable water supply, quality of services, and health—from Flint, Michigan; Cape Town, South Africa; to Delhi, India. This does not bode well for cities lacking long term strategies that respond to challenges of rising water demands for urban, agricultural, energy, and industrial production systems; and new global risks.

Ethical dilemmas and painful choices are emerging from 'lock-in' effects and environmental change to outdated governance and institutional regulatory/decision structures that have been driven by siloed systems decisions and unsustainable approaches to consistently delivering basic water, energy, food, and waste management services in an increasingly urban world. Aging infrastructure, increasingly over-allocated water resources and lack of strategies to mitigate risks (associated with urbanization, economic and environmental stress, and cybersecurity) have, in many parts of the world, brought ethical dilemmas to the forefront. In fact, this chapter outlines how many global cities are still relying on nineteenth century water policies and twentieth century infrastructure yet are now confronted with paralyzing twenty-first century challenges. In general, there is a 'rear-view mirror' approach and urban water management—continuing to look to the past to guide infrastructure planning and public policy (e.g. the loss of stationarity in water planning) with limited funds to modernize services.

2. From pain points and ethics dilemmas to urban solutions

Case studies that highlight and quantify critical pain points, narrate emerging ethical dilemmas, and outline breakthrough, interdisciplinary responses will be needed to enable new opportunities for improving quality of life, prosperity, sustainability and resilience of communities and cities. There are many examples of inaction or reaction to stresses and shocks in the urban water sector. They include current crises in water availability and quality issues in cities from Los Angeles, California to Flint, Michigan in the United States; to Sao Paolo, Brazil; Cape Town, South Africa; and Sana'a, Yemen globally; as well as increasing frequency and intensity of weather extremes and natural disasters (e.g. floods experienced in San Juan, Puerto Rico; to extreme heat in Lahore, Pakistan; to population migration in Beirut, Lebanon due to water-driven security risks in Syria; to persistent and water-related food insecurity in Addis Ababa, Ethiopia; to limited drinking water in Gaza and significant wastewater effluent discharges into the Mediterranean, increasingly leading to events of shutting down of Ashkelon's desalination plant, that now supplies up to 20% of Israel's drinking water) [6].

In this chapter, we define a spectrum of failures in current 'rear-view mirror' approaches to urban water strategy and their consequences. This is followed by the development of a preliminary urban response maturity model, in the next chapter, that's based upon learnings from recent water crisis events and other sectors (e.g., telecommunications to energy) which are also undergoing transformational change. For example, Bangladesh moving from almost zero mobile cellular subscriptions per 100 people in 2000 to over 78.8 per 100 people in 2015. In addition, solar PV prices declined almost 75% from 2010 to 2017 and onshore wind electricity by 25% to $0.06USD/kWh in 2017.

Transformational changes often come from defining vectors forward toward 'leapfrogs' in technologies decision systems and systems integration processes that move from reactive crisis response-modes, voluntary programs, and inadequate data systems to proactively accepting responsibility for informed market-enabling technologies, that could spur many new shared economy or water and energy development models—especially focused on rapidly urbanizing areas. A focus on integrated, ethics, and performance-based urban nexus strategies is defined as follows:

Urban NEXUS strategies (UNS) will refer in this chapter to an emerging approach and process that aims to integrate actors, knowledge, data, and assessment tools to inform the design of best practices that can be leveraged and shared across sectors and domains to deliver sustainable, healthy, and resilient water and energy systems and infrastructure services that improve quality of life while catalyzing urban innovation.

While the majority of this chapter explores the complex urban water challenges and responses needed ahead, an aspiration of 'leapfrogs' forward via urban nexus strategies are anticipated to help target more ambitious goals and integrated metrics for risk mitigation, to enhance a city's global competitiveness—in a rapidly evolving market place for innovative solutions to urban water crises. Trends of on-demand, data-driven analytics informing integrated, or nexus (rather than siloed)-based governance of critical resource-based services may also bring forward new ethics-driven decision and behavioral approaches as a key component to systems integration for UNS.

3. Water stressed cities and urban water-energy nexus responses

Throughout history, civilizations and cities have primarily located where water is plentiful—along coastlines, rivers, lakes, and mountains. Cities without water are a catalyst for many forms of instability—from economic and social to environmental, agricultural and political. This is an increasing challenge for many of the world's megacities, as well as smaller to mid-size cities that are urbanizing and industrializing at a rate of change that's been unparalleled in history. Between 2000 and 2025, it is expected that the number of megacities will roughly double, and with urban populations of 1 million reaching 2 million in timeframes as short as 8–12 years. This has significant implications for abilities to keep up with growth and maintain sustainable water services (**Figure 1**).

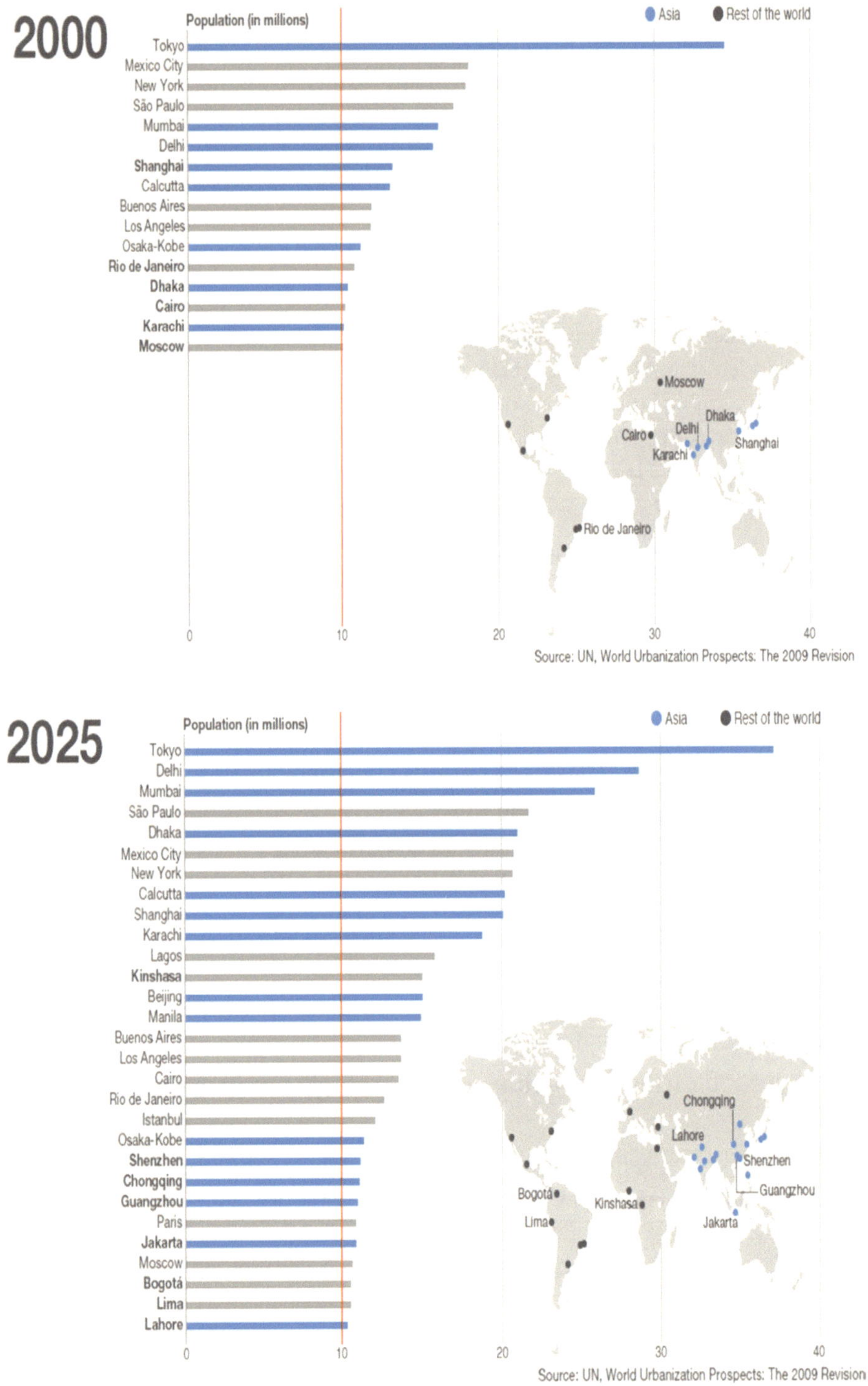

Figure 1. Mega-cities in 2000 vs. 2025 (note: today megacities represent 10% of world urban population, with smaller to mid-size cities often having more limited resources to adapt to change) [3].

http://dx.doi.org/10.5772/intechopen.82825

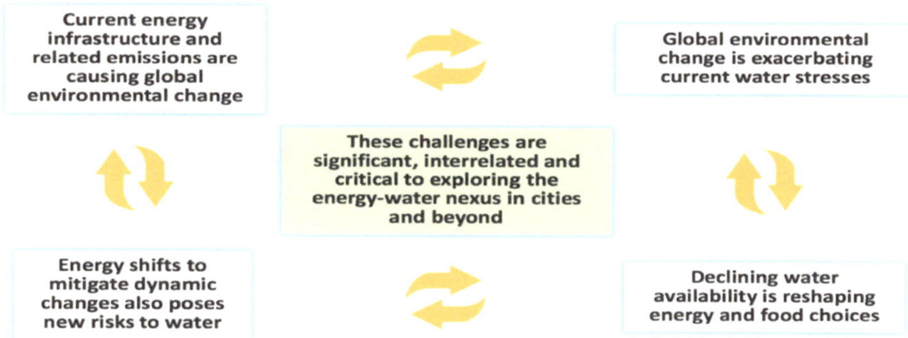

Figure 2. Interdependencies of the energy water food nexus.

Cities in wet to dry geographies are now facing increasing population and resultant resource demands. They are also in a unique position to transform water, energy and food nexus stress into strategies to vastly improve resilience and create abundance. An integrated strategy to manage nexus stress is needed for cities to thrive, economically and socially, in the twenty-first century. **Figure 2** illustrates the interactions and interdependencies of these trends and key urban resilience challenges.

4. Urban water data: understanding risks

The CDP Water program has developed research on the responses of global cities to these risks. The research indicates the cities most concerned about their water supply are in Asia and Oceania (84%), with serious risks also identified in Africa (80%) and Latin America (75%). Sixty-three percent of North American cities consider climate change a risk to water supply, with fewer cities concerned in Europe (34%). One hundred and ninety-six cities reported risks of water stress and scarcity, 132 a risk of declining water quality and 103 a risk of flooding.

CDP's new infographic report 'Who's Tackling Urban Water Challenges', produced in partnership with AECOM, the global infrastructure firm, and funded by Bloomberg Philanthropies, offers a first dataset of global water action by cities and companies. Using information gathered from 569 cities and 1432 companies, each reporting their water management activity, the database illustrates how global cities and companies are responding to the escalating challenge of resource constraints, rising demands, and changing conditions.

Businesses are also reporting to CDP on impacts from water scarcity and flooding. In 2017, 535 companies (70%) have board level oversight of water issues and are reaping the rewards, including market differentiation, shareholder confidence and business resilience. In 2017, companies committed US$23.4 billion across more than 1000 projects to tackle water risks across 91 countries worldwide, including desalination, reclaiming waste-water and improving irrigation to avoid droughts [4].

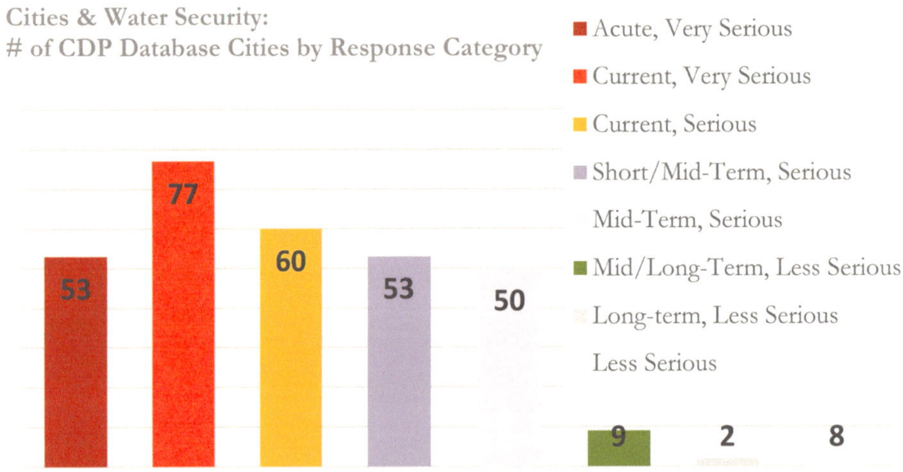

Figure 3. Urban risk timescales, magnitude, CDP cities water security database [5] (n = 312 cities; self-reported).

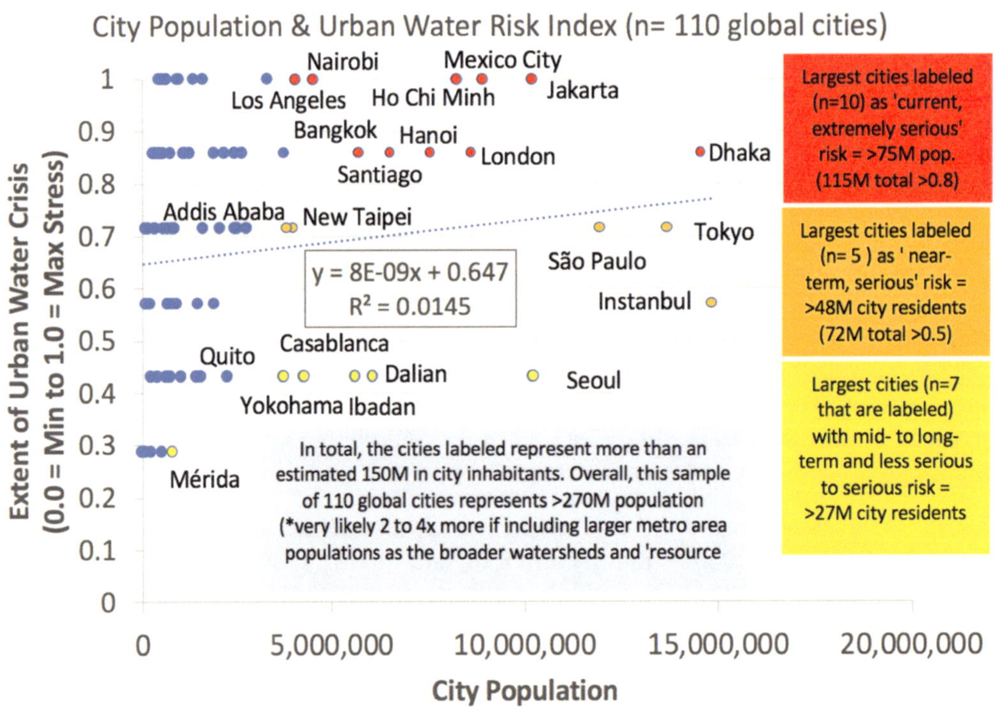

Figure 4. One hundred and ten global cities that are categorized into lower, medium and higher water risk categories. Re-analysis of September 2017 data from CDP [5].

Data re-analyses of the CDP Cities and Water Security self-reported database are shown in **Figure 3**. This includes exploring data from 312 cities in terms of water insecurity risk type, level of severity and the temporal nature of these risks (keeping in mind a need to balance the biases that may be associated with self-reported data). This is followed by a greater emphasis

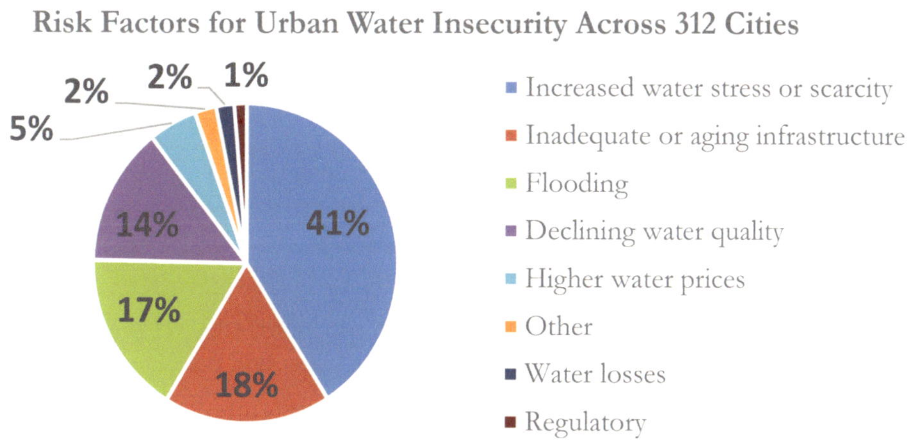

Figure 5. Proportional assessment of types of risks shaping water insecurity [5].

in next section on solutions, decision systems, and maturity levels—to operationalize the framework—on urban water strategy maturity, focusing on systems integration and service innovation across sectors and 'siloes' (**Figures 4** and **5**).

5. Exploring risk descriptions from 'hotspot' cities

An overview of impacts in the cities with some of the highest risk of water insecurity are provided below.

Jakarta

- Scarcity of clean water, especially in areas near the coast of North Jakarta and West Jakarta
- Pollution caused by industrial and household activities
- Increased frequency of rain affecting the area, with increased inundation/flood areas

Karachi

- Water riots expected as 1 Billion gallon per day requirement is only met with 600 Mgd

Nairobi

- Currently, the output of all the water sources for the city is 570,000 cubic meters against a demand of 740,000 cubic meters. This means we only meet 77% of current water demand.
- The current reticulation infrastructure is very old and, as a result, contributes to huge water losses due to leakages.
- With the flooding associated with the heavy downpours that the city experiences, there have been instances where water pipes have been swept away.

Moscow

- Risk of release of hazardous polluting substances, leading to failure of technological modes of sewerage networks and sewage treatment plants.
- Risk of accidents on sewerage networks, pumping stations and wastewater treatment plants in connection with wear and insufficient volume of measures for their renovation, as well as in connection with failure of external power supply.
- Risk of accidental pollution of water sources, existing due to anthropogenic pressures, leading in particular to deterioration of water quality, primarily on organoleptic and microbiological indicators, the content of organic substances and petroleum products.
- Mass development of cottages development in water-collecting area and discharge of untreated waste water lead to gradual degradation of small rivers, deterioration of self-purification capacities of water bodies, and algal blooms. Deterioration of water supply systems also affects the quality of the water supplied to consumers.

Cape Town

- The city has generally been able to successfully manage and reduce demand growth; however, Cape Town is currently suffering from a drought lasting several years (currently in year 3) which has severely impacted the City's water storage.
- Stringent level 3b water restrictions have been put in place to reduce demand further. It is anticipated that in the longer term, water demand will continue to grow and place stress on the supply system.
- The city is currently conducting cooperative planning with the national Department of Water and Sanitation to ensure that additional water supply infrastructure is constructed to avoid a long-term water deficit in the region. Climate change is expected to change rainfall patterns, and this has been included as a scenario in the planning for future infrastructure. Climate change is expected to reduce rainfall, increase evaporation and increase demand due to increased temperature.

Mazabuka

- Inadequate rainfall results in droughts, damaged crops, declining wildlife and deaths
- Hydroelectricity issues, for instance—no electricity in homes at certain intervals due to road shading and you only see power for only 2 hours.
- People consuming contaminated water, due to dried and stagnant water sources, leads to water borne diseases.

Mexico City

- The city has identified and mapped flooding impacts for vulnerable areas—targeting 5.6 million inhabitants.

- So far there are over 41.1% leaks in the water system. There is constant pressure on the aquifer and the 2040 water plan has identified this as the highest level of risk.

Lagos

- Wastewater treatment plants available are not sufficient to treat wastewater for the whole city
- Over abstraction of ground water from aquifers
- Mismanagement of water resources, such as leaking taps, lead to scarcity of water availability
- Reduction in quantity of water available for domestic, industrial and commercial use

Johannesburg

- A crippling drought that was associated with El Niño severely affected South Africa's summer rainfall in regions including Johannesburg.
- Dam levels were at critical low levels such that water restrictions had to be imposed and penalties on those who used excessive water.

Los Angeles

- Deferred maintenance on water system

Las Vegas

- As Lake Mead's level declines, concerns of declining water quality due to increased salinity.
- Recent third intake project at lowest part of lake to mitigate water quality concerns.
- Water restrictions; yet if Lake Mead continues to decline, a Federally mandated cut in Southern Nevada's water allocation may occur, and nearly did during 2015-2016.

6. Extent, consequences, and drivers of urban water crises

This section focuses on city and regional-decision making processes that will matter. Below offers example 'cities most likely to run out of drinking water', with some related trends, the potential consequences in these changing urban environments, and multi-level (e.g. city, state, national) responses. While data integration across systems may help to inform future risk communication, increased availability of data and a need for data integration and transparency requires new analytics coupled with decision processes, enabling a higher level of maturity in response levels that move from reactive to proactive. Several predictive factors are therefore noted from both data and literature-reviewed in these cities, offering contexts where scarcity may increase to insights on where 'abundance strategies' may prove of value (see Sarni and Sperling).

6.1. Sao Paolo (estimated metro region population: 21 million)

- *Indicators on extent of crisis:* During the 2014 water crisis, the city's main reservoir was at 3% capacity and the City had less than 20 days' waters supply [7]. In the period of early twentieth century up to 2015, 12-month estimates of rainfall reached levels that were at half the amount of all previous worst 12-month periods [8].
- *Crisis consequences:* $925 million of investment by water company, Sabesp, in three new water pumping projects to provide enough back-up water to survive a drought similar to the 2014–2015 event; many taps flowed for only a few hours every 4 days; theft and looting of emergency water trucks; 71% of city population experienced problems with the water supply during worst month, and with most acute impacts of going dry felt in the *Periferia*—poorer districts on the city; while wealthier residents built water tanks and purchased water from private sources, outlying cities saw large protests with some turning violent when city tried to cut off from water network entirely; dehydration of children, women with urinary problems from not drinking enough water, and significant business risks to factories and farm output; water shortages also impacted hydropower plants in the metro region forcing energy rationing (with principal hydropower reservoirs at 17% capacity)
- *Risk factors and Responses:* 10-fold increase in city population from 1950 to 2005 and uncontrolled urban expansion with informal settlements lacking adequate water services; pollution of rivers, deforestation in Amazon River basin and water-intensive agriculture; network leakages leading to 25% of produced water not reaching water users; main responses to date have included expensive, supply-side engineering / infrastructure investment to avoid future shortages rather than reducing consumption and leakage.

6.2. Bangalore (estimated metro region population: 11 million)

- *Indicators on extent of crisis*: City water distribution networks only cover the central area of the city, whereas surrounding areas are not connected and instead get their water supply from tanker trunks (typically relying on quickly shrinking groundwater supply that have dropped from depths of 150–200 ft. to 100 ft. or more in some places).
- *Crisis consequences:* Sept 2, 2016 Supreme Court issued-order for Karnataka to release extra water (10,000 cubic feet of water per second, then 15,000 from Sept 5 to 15 Sep and 12,000 until Sept 20) from the Cauvery river to ease a shortage threatening crops in Tamil Nadu; this led to violent/deadly protests in Bangalore forcing closures of hundreds of companies and public transit system (city police imposed an emergency law prohibiting public gatherings, with more than 15,000 officers deployed across the city).
- *Risk factors and Responses:* Total extraction wells from 5000 to 450,000 in the past 30 years; groundwater table drop from 10-12 meters to about 76–91 meters in just two decades with minimal groundwater recharge and rising water body pollution due to unplanned urbanization. The Bangalore Water Supply and Sewerage Board (BWSSB) is working with the Japanese International Cooperation Agency to divert 10 thousand million cubic feet of

water from the Cauvery River for drinking water (anticipated to provide ~100 L of water per person per day), gaining an additional 50% more than current supply. Other responses to date have included plans for an 18 month project to divert water from another river, the Netravati, as well as mandatory construction of rainwater harvesting facilities (collecting $300,000 in fines per month from those not complying) and recycling water using sewage treatment plants [9, 10].

6.3. Beijing (estimated metro region population: 22 million)

- *Indicators on extent of crisis:* 100 cubic meters is available per person per year (note: less than 1000 cubic meters/capita annually is considered "water scarce" by UN standards). Price of water remains a quarter of the world's average; 12 consecutive years of drought, noted in 2011, led to investments in desalination and piping from the Bohai sea [11].
- *Crisis consequences:* Three new routes for diverting water from south to north could cost more than $80 billion with experts having doubts as to this being a long term sustainable solution (yet rather a 'lifeline' of water in the short term) [12]; 112 million regional population across Beijing/Tianjin/Hebei faces half of country's acute scarcity, where 28,000 rivers have disappeared in past 25 years, groundwater is falling by up to 1–3 m a year, and some parts of Beijing subsiding by 11 cm a year. Yellow River water supplies millions, yet is now at a tenth of 1940 flow levels and often fails to reach the sea [13].
- *Risk factors and Responses:* In 2017, 8.8% of water was unfit even for agricultural or industrial use with pollution causing further risk to supply [13]; estimates of more than 50 million people in Beijing by 2050 [14]; groundwater decline and widespread water pollution; diverting water from Yangtze river in south; 'sponge city' pilots using up to 70% of rainfall; water recycling; drought resistant crops; Tianjin desalination infrastructure investments

6.4. Cairo (estimated metro region population: 19.5 million)

- *Indicators on extent of crisis:* Population growth (Egypt's overall population is expected to double by 2050) and significant environmental pollution (especially chemically treated sewage disposal and industrial waste that's killing crops) have led to the Nile river becoming the recipient of significant urban wastewater due to the lack of wastewater treatment plants in Cairo and rural agriculture and industrial runoff [15]. Below offer results from water samples at different water treatment plants in Cairo in the past decade [16]: cryptosporidium was found in 50% of samples taken from the Fowa drinking water treatment plant to 100% of samples in the El Nomros plant, and with Giardia as high as 33% in the El Hawamdia to 50% in Meet Fares. Over the past decade, the peri-urban areas, or outskirts of Cairo, have also been under significant development—this has included illegally constructed buildings linked to unauthorized use of primarily leaky water pipelines that then waste limited urban water supplies.

- *Crisis consequences:* Some have noted the Nile is now running out of clean water and with continued uneven water distribution, misuse of water resources, and inefficient irrigation techniques – 20 cubic meters per person of internal renewable freshwater resources has now become the norm nationwide. By 2020, Egypt will consume 20% more water than is available. This has meant an annual 7 billion cubic meters water deficit and UN predictions that the entire nation of Egypt could run out of water by the year 2025 [17]. In addition, due to groundwater uses for irrigation affecting the water table, structural integrity of several buildings and historic monuments in Cairo remain at risk as well.
- *Risk Factors and Responses:* Serious urban health hazards ranging from diarrhea, eye infections and rheumatism – associated with exposure to sewage – have been indicators of increasing risk since the late 1970s. This includes a high ranking for the number of deaths related to water pollution. In response, USAID alone has invested more than $3.5 billion to improve water and sanitation services for over 25 million Egyptians (in a country just under 100 million today and expected to reach 200 million in next 50 years). More recently, in West Cairo, a $727 million project was developed and implemented to improve wastewater collection, treatment, and disposal [18].

6.5. Jakarta (estimated metro region population: 30 million)

- *Indicators on Extent of Crisis*: Excessive groundwater water extraction over the last three decades has led the city to be one of the fastest sinking cities in the world and people are leaving [19]. Land subsidence is at a rate of 3 cm to 20 cm per year in parts of city (including 5 to 8 cm a year in the northern half of the city). Meanwhile, Jakarta's population has increased from ~8.3 million (2000) to 10,075 million (2015), without increases in environmental service (e.g. clean water provision, waste water treatment) capacities. Flooding occurs almost every time it rains more than three hours (with widespread flooding inundating up to 40% of the city in 1996, 2002 and 2007 [20]), only 50% of households have piped water (with new connections for low income households remaining very low – 85% of households that have connections fall into tariff categories of middle class or above [21], and the city now produces more wastewater than clean tap water. Finally, river water quality status in Jakarta has reached levels of 81% indicated as highly polluted in 2004 up to 94% in 2007, while green space in Jakarta's 1985–2005 to 2000–2010 spatial plans decreased from 26.1% to 13.94%.
- *Crisis Consequences:* Only 4% of housing in Jakarta has wastewater treatment plant connections. Jakarta has become a heavily polluted city in terms of sewage and water. This includes 70% of waterways being blocked, as a central driver of the city's chronic flooding problems. Twenty percent of daily waste still ends up in local rivers and canals, causing significant illness. Flooding has also led to the displacement of more than one million people, and billions of dollars in losses have ensued. In 2007 alone, nearly 70% of the city was submerged by floodwaters – 52 fatalities and displacement of more than 450,000 [22].
- *Risk Factors and Responses:* With support of international donors, and national government, the city administration started in 2016 to dredge its 17 rivers and canals for the first time since the 1970s. Other (somewhat inadequate) responses have included building of

shopping malls and apartment blocks as unchecked development, that would normally be retention ponds, swampland and other open spaces that would normally absorb rainwater.

6.6. Moscow (Est. 12 million)

- *Indicators on Extent of Crisis*: Insecure sources of water; water stress; water pollution- 56% of water supply sources fail to meet safety standards [23]
- *Crisis Consequences:* Health risks due to heavy metals, soil and groundwater contamination
- *Risk Factors and Responses:* Inadequate or aging infrastructure, declining water quality (CDP, 2017); and low incentives to take steps to improve quality of wastewater.

6.7. Istanbul (Est. 14 million)

- *Indicators of Extent of Crisis:* Distant from water sources; from 2006 to 2008, rainfall was lowest recorded in last 50 years; in 2014, another drought led to reservoirs dropping to as low as 29%; treatment plants are outdated; supply deficit (in million m3/year) was 473.2 in 2005 and 682.0 in 2010 [24]
- *Crisis Consequences*: Ongoing pollution to waterways; from drought to mega infrastructure projects of dams and canals; ecosystem impacts to nearby streams, water tables, large costs, and potential flooding anticipated with heavy rains.

  Risk Factors and Responses: Increased water stress or scarcity (as serious medium-term risk); 2.8% annual growth rate (population doubling in 25-year period); Melen dam now expected to help Istanbul meet water demand until 2071, yet with higher water prices in long-term; and continued prolongation of dry season creating pressure on water resources (CDP, 2017).

The similarities among these examples are: 1) continued lack of data and metrics on water use and management; 2) lack of incentives or markets for solutions; 3) aging, centralized systems for infrastructure; 4) lack of business models for small, modular systems that may increase abundance and resilience; and 5) inadequate institutional mapping of roles to inform accountability/transparency in crisis response. These examples offer a diverse range of the sets of challenges, consequences, and emerging risks and responses. **Table 1** offers another summary of literature review approach, comparing Mexico City, London, Tokyo, and Miami, on recent crises related to resource quantity, quality, supply, demand, equity, and choices. These examples present a picture of the grand urban water challenges of the 21st century, with both differences and similarities between cities. Proactive strategies and integrated responses focused on the growing number of cities at risk as frontlines for innovation may continue to emerge. These examples also motivate questions for ongoing exploring of long-term impacts, using data to generate understanding on how best to help reduce costs, improve water security, modernize infrastructure assets, build resilience and ensure sustainable revenue models.

City	Resource quantity	Resource quality	Supply/demand challenges	Distributional & procedural equity	'Painful' choices and resilience priorities
Mexico City	Declining water resources; growing population; groundwater dependence limiting aquifer recharge	Poor water quality adaptations: increased purchasing of bottled water; plastic contributing to flood and drainage risks	Water scarcity; uncontrolled urban expansion; 30–40% of water supply lost to leaky systems & 1000s of informal taps	Water resources transferred over long distances, connecting city populations to resources in distant places; prioritization processes; ongoing flood risk	Investments in expensive water storage/supply-side water-management strategies reliance; new deep wells / water transfers from state of Veracruz; reuse strategies only recently emerging
London	London has a high level of leakage with around a quarter of London's water lost	Increased population/ heavy rainfall to further burden a drainage system that is in places already at capacity	Half of London's water mains pipe infrastructure is over 100 years old and a third is over a 150 years old	14,000 properties are at high risk (0.33% annual probability) of fluvial flooding and 140,000 properties are at high risk of surface water flooding	London flooding in 2000, 01, 03, 06, 07, 14 and 16. Infrastructure upgrading & reducing vulnerabilities to tidal, river, surface water, groundwater and sewer flooding
Tokyo	80% from Tonegawa & Arakawa River; 20% Tama River	Declining water quality	Increased water stress	Inadequate and aging infrastructure	Risks of flooding; investments in infrastructure and resource efficiency gains
Miami	Built on Biscayne aquifer	Contamination by seawater threatens quality of Miami water supplies	Increased water stress and scarcity	Water table overwhelming septic systems	Flooding risk and toxic chemicals on Superfund and other industrial sites into aquifer; 'desalination may one day be Miami's only option.'

Table 1. Comparisons of cities, megatrends, and resilience priorities.

7. Moving from grand challenges to critical nexus opportunities

New paradigms are emerging with water, energy, and related technologies identified as grand challenges. Innovation can create opportunities to both better leverage data and meet a global need for safe, secure and affordable water and through higher degrees of systems integration between water, energy, and 'X' systems (X is defined as a variable that may include food, land, waste resource recovery, environmental, economic, and urban systems – depending on local context priorities). Building on the National/Global Academies of Engineering twenty-first century grand challenges of 'restore and improve urban infrastructure', 'secure cyberspace', and 'provide access to clean water', this section of the chapter evaluates the critical opportunities to increase water, energy, environmental, social, economic, and security benefits, while significantly reducing costs, societal and business risks (e.g. supply chain vulnerabilities, conflict, economic outputs exposed and vulnerabilities to multiple hazards).

Given uncertainties associated with interrelated challenges, it's critical to not only define the grand water security challenge(s) for cities in the twenty-first century, yet further build up a more robust evidence base for integrated solutions. For purposes of defining urban community challenges in this chapter, we start with the fact that one in four of the world's 500 largest cities are already in a situation of "water stress" (ref, 2014) and that water crises, since 2011, have consistently been ranked in the top five of global risks in terms of impact (WEF Global Risks, 2018).

The economic consequences for inaction and holding on to old technology solutions are clear. The World Bank Report High and Dry: Climate Change, Water and the Economy, lays out a grim vision of inaction. Below are a few of the conclusions from the report:

- "Water scarcity, exacerbated by climate change, could cost some regions up to 6% of their GDP, spur migration, and spark conflict.
- The combined effects of growing populations, rising incomes, and expanding cities will see demand for water rising exponentially, while supply becomes more erratic and uncertain.
- Unless action is taken soon, water will become scarce in regions where it is currently abundant—such as Central Africa and East Asia— and scarcity will greatly worsen in regions where water is already in short supply—such as the Middle East and the Sahel in Africa. These regions could see their growth rates decline by as much as 6% of GDP by 2050 due to water-related impacts on agriculture, health, and incomes.
- Water insecurity could multiply the risk of conflict. Food price spikes caused by droughts can inflame latent conflicts and drive migration. Where economic growth is impacted by rainfall, episodes of droughts and floods have generated waves of migration and spikes in violence within countries."

The report also maps out the benefits of addressing the water crisis, such as improved economic development, and a call for action in improving agricultural water efficiency, better planning and investments in infrastructure to ensure more secure water supplies and availability. Most recently, a report released by NASA illustrates the impact of unstainable pumping of aquifers. According to NASA, "The world's largest underground aquifers—a source of fresh water for hundreds of millions of people—are being depleted at alarming rates, according to new NASA satellite data that provides the most detailed picture yet of vital water reserves hidden under the Earth's surface. Twenty-one of the world's 37 largest aquifers—in locations from India and China to the United States and France—have passed their sustainability tipping points, meaning more water was removed than replaced during the decade-long study period, researchers announced Tuesday. Thirteen aquifers declined at rates that put them into the most troubled category. The researchers said this indicated a long-term problem that's likely to worsen as reliance on aquifers grows."

Population growth is also placing significant stress on energy and food production, which is further exacerbated by water scarcity. The global population is currently increasing by approximately 70 million people each year. As a result, the total global population is projected to reach 9.6 billion by the year 2050 [25]. The International Union for Conservation of Nature (IUCN) estimates that by 2050, water, energy, and food demands will increase by 55, 80, and 60%, respectively [26].

This growth will increase the pressure on limited water, energy, and food resources. Energy consumption is estimated to increase by 1.6% each year, amounting to an increase of about 36% by the year 2030. Additionally, pressure on agricultural resources will increase through societal habits such as consumption of more livestock and vegetable oils. The number of calories that a person ingests each day is expected to increase from 2373 kcal/person/day in 1969/1971 to 3070 kcal/person/day in 2050. Urbanization will yield more industrialization and water usage, and water demand will increase globally from 4,500 billion cubic meters to 6,900 billion cubic meters by the year 2030. This estimation assumes that the efficiency in water technologies does not improve, and the projected demand is about 40% over our currently accessible and reliable supply [27]. These challenges demonstrate the need for data, nexus solutions, and the combining of emerging solution pathways for technology and services, regional planning, policy and governance, and new behaviors and decisions if urbanization is to be steered toward a sustainable trajectory.

In conclusion, a new set of water and energy ethics are needed to maximize human and ecosystem health and prosperity. Bringing systems together via urban nexus strategies—that amplify synergies, reduce tradeoffs, and can transition resources from scarcity to abundance—will be foundational to a more resilient human, natural, to cyber-physical system that supports diverse activities today and for future generations. As noted, multiple risks, vulnerabilities, and early signs of stress abound. Therefore, the abilities and capacities to harmonize human to ecological needs will require new, integrated ways of using and managing water, energy and other systems and services. Elegant designs will emerge soon, from crises or proactive actions in urban contexts.

Author details

Josh Sperling[1*] and Will Sarni[2]

*Address all correspondence to: joshua.sperling@nrel.gov

1 National Renewable Energy Lab, Denver, Colorado, United States

2 Water Foundry, Denver, Colorado, United States

References

[1] Venkatesh S, Sengupta S, editors. 36 Percent of Cities to Face Water Crisis by 2050. Center for Science and Environment; 2018. Available from: www.downtoearth.org.in/news/36-per-cent-cities-to-face-water-crisis-by-2050-59985 [Accessed: July 2018]

[2] Distel O. Water—A National Priority. Israel: NewTech; 2017. Available from: https://www.youtube.com/watch?v=1skF7jJmbXw [Accessed: July 2018]

[3] UN. World Urbanization Prospects; 2009 Revision

[4] https://www.cdp.net/en/research/global-reports/global-water-report-2017

[5] CDP. 2017. Available from: https://data.cdp.net/Cities/2017-Cities-Water-Risks [Accessed: 2018]

[6] BBC. The 11 cities most likely to run out of drinking water – Like Cape Town. 2018. Available from: https://www.bbc.com/news/world-42982959 [Accessed: July 2018]

[7] Ritter K. São Paulo Heading To Another Dry Spell. Circle of Blue. 2018. Available from: https://www.circleofblue.org/2018/water-climate/drought/sao-paulo-heading-to-another-dry-spell/ [Accessed: July 2018]

[8] Watts, J. 2017. The Amazon Effect: How Deforestation is Starving São Paulo of Water. Available from: https://www.theguardian.com/cities/2017/nov/28/sao-paulo-water-amazon-deforestation [Accessed: July 2018]

[9] BBC. Is India's Bangalore Doomed to be the Next Cape Town? 2018. Available from: https://www.bbc.com/news/world-asia-india-43252435 [Accessed: July 2018]

[10] eNCA. Bangalore Faces Man-Made Water Crisis. 2018. Available from: https://www.enca.com/world/bangalore-faces-man-made-water-crisis [Accessed: July 2018]

[11] Qiujian S, Jei F. How to Quench a Thirst? Beijing Is Desperately Short of Water – And Uncertain how to Fix the Problem, Torn between Desalination and Pumping Resources from the South. China Dialogue. 2011. Available from: https://www.chinadialogue.net/article/show/single/en/4396-How-to-quench-a-thirst

[12] Westcott B, Wang S. Can China Fix its Mammoth Water Crisis before it's Too Late? CNN; 2017 Available from: https://www.cnn.com/2017/03/21/asia/china-water-crisis/index.html

[13] Parton C. China's acute water shortage imperils economic future. Financial Times. Available from: www.ft.com/content/3ee05452-1801-11e8-9376-4a6390addb44 [Accessed: July 2018]

[14] Thompson F. Modeling Beijing's water crisis. Eos. 2017;**98**. DOI: 10.1029/2017EO085211

[15] Agrama A. Assessment of surface water quality in Egypt. Journal of Applied Sciences Research. 2012;**8**(4):1944-1951

[16] Ali MA, Al-Herrawy AZ, El-Hawaary S. Detection of enteric viruses, Giardia and Cryptosporidium in two different types of drinking water treatment facilities. Water Research. 2004;**38**(2004):3931-3939

[17] Dakkak A. Egypt's Water Crisis–Recipe for Disaster. EcoMena; 2017. Available from: https://www.ecomena.org/egypt-water/ [Accessed: July 2018]

[18] USAID. Egypt: Water and Sanitation. 2018. Available from: https://www.usaid.gov/egypt/water-and-sanitation [Accessed: July 2018]

[19] Kusumawijaya M. Jakarta at 30 million: My city is choking and sinking—It needs a new plan B. 2016. Available from: https://www.theguardian.com/cities/2016/nov/21/jakarta-indonesia-30-million-sinking-future

[20] Apip, Sagala S, Pingping L. Overview of Jakarta water-related environmental challenges. In: Water and Urban Initiative Working Paper Series. United Nations University; 2015. Available from: unu.edu/research/water-and-urban-initiative.html and http://collections.unu.edu/eserv/UNU:2872/WUI_WP4.pdf

[21] UNDP. Jakarta Indonesia: Case Study (Water). Special Unit for South-South Cooperation; 2012. https://www.esc-pau.fr/ppp/documents/featured_projects/indonesia.pdf

[22] Cochrane J. What's clogging Jakarta's waterways? You name it. New York Times: Jakarta Journal. 2016. Available from: https://www.nytimes.com/2016/10/04/world/asia/jakarta-indonesia-canals.html

[23] Ritter K. Pollutants and Heavy Metals Taint Moscow's Water Supply. 2018. Available from: https://www.circleofblue.org/2018/europe/pollutants-and-heavy-metals-taint-moscows-water-supply/ [Accessed: July 2018]

[24] Bekiroğlu S, Eker O. The importance of forests in a sustainable supply of drinking water: Istanbul example. African Journal of Agricultural Research. 2011;**6**(7):1794-1801. Available from: www.eip-water.eu/sites/default/files/Istanbul%202016%20EDS.pdf [Accessed: July 2018]

[25] Sarni W. Deflecting the Scarcity Trajectory: Innovation at the Water, Energy, and Food Nexus. Deloitte University Press; 2015

[26] International Union for Conservation of Nature. The water-food-energy nexus: Discussing solutions in Nairobi. 2013

[27] Sarni W. Beyond the Energy – Water – Food Nexus: New Strategies for 21st Century Growth. Dō Sustainability; 2015

Chapter 3

A Call to Cities: Run Out of Water or Create Resilience and Abundance?

Will Sarni and Josh Sperling

Additional information is available at the end of the chapter

http://dx.doi.org/10.5772/intechopen.82853

Abstract

New management choices, with new approaches to urbanization and integrated water-energy-food management, are emerging as critical to combat water stress. Urban strategies and tactics are explored in this chapter with a focus on scaling effective solutions and approaches. This includes a focus on small, modular, and integrated water-energy-food hubs; off-grid and localized "circular economy" services that are affordable, accessible, and reliable; blended finance for new technologies, infrastructure and business models, strategic plans, and policies; and urban, behavioral, and decision sciences-informed decisions and new public-private-research-driven partnerships and processes. There are two key messages: first, business as usual could lead to "running out" of water where it's needed most—in cities and for agricultural and industrial production. Second, "innovators" and "early adopters" of market-based and data-driven efforts can help scale solutions led by people and communities investing in new ways to integrate urban water, energy, and food systems. The chapter concludes with discussion on a new, proactive "maturity" model, enabling integrated urban infrastructure systems, governance, and cross-sector innovation. This includes market-based and data-driven responses that first focus on improving quality of life, sustainability, and resilience of communities, bringing valued services via water-energy-food nexus decisions.

Keywords: water, energy, infrastructure, market solutions, nexus governance, cities

1. Introduction

For the first time in decades, water, energy, food, and other systems are experiencing significant innovations. These innovations—with breakthroughs in distributed, modular, and digitally connected technology and governance strategies for "energy and water as a service" options—are all recognized as having potential to transform dynamics that may lead toward dramatically different futures. One future is more water and energy use, aging infrastructure with deferred maintenance, and unhealthy communities. The other future is significant public and private benefits of more water choices, greater affordability and accessibility, and healthier, more livable communities, with less energy use and fewer costs. A new ethic around water is needed to achieve the latter. As this resource becomes "precious," new levels of maturity for co-designing urban nexus experimentation for resilience and abundance may emerge. Furthermore, market-based approaches and structural shifts in governance using emerging data, by bringing new transparency, will inform new ethical behaviors and moral decisions for crises and upscaling of innovations. Cities, utilities, and service providers will all be on frontlines of integrated solutions for the complex water dilemmas ahead. However, predicting and paying attention to the "late adopters and laggards" may be equally key.

Business as usual has not led to sustainable, healthy, nor resilient future pathways for urban and rural communities. However, necessity has been noted as a unique catalyst for innovation and may prove a key motivator for new approaches. In particular, cities are rapidly growing demand centers for maturity in sustainable and integrated resource management. Unique pressures and risks to economic growth [1], development, and social and ecosystem well-being are motivating many global cities currently struggling to supply equitable access to services to invest more readily in safe, secure, and affordable water systems. Lessons across urban systems, including those with reliable, affordable, secure water and energy, offer a key opportunity to improve management of impacts from pollution and extreme weather events.

Today, when supply exceeds demand, the absence of public policy or poorly conceived strategies for water can often go unnoticed with few, if any, consequences. This is no longer an option in a world where increased demand is creating scarcity and difficult allocation choices. In many parts of the world, demand currently exceeds supply due to rapid increases in population and associated needs for energy, food, and products. As a result, a key need is to accelerate urban water-energy nexus technology, partnerships, financing, business models, and public policies to ensure adequate water and energy infrastructure systems and governance approaches to upgrade urban services enabling economic development, thriving environments, and social well-being.

Data

- How can research-practitioner communities advance data-driven decisions—what strategies can best generate enhanced data and models of increasing equity to resilience?
- What nexus strategies will encourage innovation and reinventing of global urban water systems across emerging/converging energy-water-food systems/services environments?
- What will be the future of revenue and infrastructure (re)development, and in what direction do we need to move for harnessing technology and new services for positive economic/business model outcomes while reducing financial or environmental risks?

Assessment methods

- What are key priorities, modernization definitions, and related nexus metrics for reporting (e.g., water productivity for energy and agriculture systems or energy productivity for water and wastewater treatment to food storage and distribution)?

http://dx.doi.org/10.5772/intechopen.82853

- What input data, models, and information resources can be drawn upon?
- How best to integrate resilient infrastructure investments and decision support system tools that are cross-sector, multilevel, and enabling of cost-effective/sustainable PPPs that offer market-based and data-driven strategies?

Box 1. Key questions on long-term impacts and urban nexus transformations.

While water-related crises are unfolding in cities all over the world, with less than adequate response strategies [2, 3], many will agree that cities will likely not "run out" of water; instead, these crises will lead to difficult and painful choices on allocation. In the long-term, what water may be left could primarily go to the privileged. This is why the questions noted in **Box 1** may become increasingly critical.

Cape Town offers one illustrative example: "About a quarter of Cape Town's population lives in the informal settlements, where they get water from communal taps instead of individual taps at home…." "One reality is that those 1 million people out of a population of 4 (million) only use 4.5 per cent of the water."—http://www.cbc.ca/news/world/cape-town-water-day-zero-1.4518226.

2. Data and insights informing a city maturity framework

How did Cape Town get to the point where it had to plan for "Day Zero," when extreme restrictions on access to water will be enforced? Some might ask, are there not ample technological solutions to address the shortages? How does a metropolitan area of nearly 4 million people manage to become "the first major city in the modern era to face the threat of running out of drinking water"? What we do know is that Cape Town is not alone, and the challenges extend beyond technology to multiple institutional challenges and the way water is valued today. With a recent report exposing 11 other cities at risk, many anticipate significant underestimates as to the number of cities in peril (https://www.greenbiz.com/article/avoiding-next-cape-town-water-strategy-shared-responsibility).

First, let us explore the narratives for Cape Town. What went wrong? Where should blame be placed, and which institutions and new policies led to a city more equipped for creating innovative responses? In addition, how are such problems prevented from arising again as access to fresh water becomes one of dominant public policy issues of the twenty-first century? Barriers identified to date have included:

- Lack of awareness
- Lack of sustainable business models
- Lack of dynamic and relevant pricing policies
- Lack of clear legal frameworks
- An abundance of conservativeness and reactive approaches

While additional, local, context-specific factors could continue to drive future water crises in Cape Town and other cities, these non-technological issues offer a much broader set of risks faced across many cities. Whereas drought and water supply shortage help to demonstrate the central role water plays in our lives, this role seems to remain underappreciated and, as a result, undervalued.

On the surface, the underlying story is about a failure in how the public sector manages water. All too often, public water policy is based upon poor data and information, inadequate public engagement, and a belief that the past is a good guide to the future. In other words, water is too often treated as a taken-for-granted asset rather than a strategic resource for economic development, social well-being, ecosystem health, and competitive advantage. Where water is viewed strategically, e.g., Israel and Singapore, water scarcity and stress do not limit economic development and business growth yet enhance it as these countries often turn their water technology innovation investments into an export initiative (e.g., WATEC, Singapore World Water Week).

In response, revolutions and leapfrogs in approaches may be needed to ensure both bottom-up to top-down approaches are accelerated to hold cities, states, utilities, businesses, communities, and national leaders accountable for this shared resource, both to each other and to future generations counting on getting this right for coupled human-environmental-economic security and regional stability [4]. Additional illustration of city examples representing inaction, reaction, and painful choices will be needed so we can learn from past failures and enable new advances and water culture shifts. Likewise, cities moving to resilience and abundance, through a new set of values, ethics, and norms—coupling technology-planning-governance-behavior-finance systems need to be further explored. This is elaborated on in the following sections.

3. Integrated approaches in moving to resilience and abundance

While there is considerable discussion of the energy-water-food nexus and associated impacts, there is less of a focus on innovative solutions. Resource stress and scarcity foster innovation in technologies, financing, business models, and partnerships. We are also seeing innovation in public policy to address nexus stress and scarcity. Public policy innovation is catching up to advances in technology, financing/funding, business models, and partnerships. Collectively, innovation will move the world from scarcity to abundance if managed effectively by the public sector, companies, nongovernmental organizations, and civil society. This section of the chapter explores initial innovative approaches to addressing water security within a context of energy, water, food, land, climate, and other systems and services that help reframe decisions as integrated solutions to *create abundance.* (https://www.routledge.com/Water-Stewardship-and-Business-Value-Creating-Abundance-from-Scarcity/Sarni-Grant-Orr/p/book/9781138642553 and https://www.routledge.com/Creating-21st-Century-Abundance-through-Public-Policy-Innovation-Moving/Sarni-Koch/p/book/9781783537518).

Also, invaluable experience was gained in working with XPRIZE (www.xprize.org), Imagine H2O (www.imagineh20.org), and 101,010 (www.101010.net) along with multinationals and nongovernmental organizations (NGOs) on water risks to food and energy production.

The bridging between engineering, entrepreneurship, data science, and public policy, among other fields, has the potential to help chart a better path forward. In this chapter, we refer to a new field of urban nexus science and innovation, one that helps move toward recognizing that 9 billion people deserve access to energy, food, safe water, sanitation and hygiene as part of a fundamental quality life, with a critical focus toward helping inform the integrated design and scaling of innovative solutions, while not just continuing to rely upon current 'siloed' approaches, which have yielded incremental progress.

3.1. An urban nexus response maturity model: from incremental transitions to breakthrough (or leapfrog) transformations enabled by market insights

Below offers a framework toward nexus-based solutions that enable increased abundance. Markets setting ambitious goals and using new performance metrics for risk mitigation are enabling new city competitiveness opportunities. Market-based approaches and structural shifts in governance using nexus data are also informing behaviors and decisions that may offer opportunities to upscale strategic innovations.

Cities, utilities, and service providers are increasingly faced with complex and challenging water-related dilemmas. **Figure 1** offers examples of how different types of urban water actors and institutions could begin to respond in the context of their own water systems and services, to think about linkages and opportunities by also tackling related food and energy systems and services.

This system integration framework and future mapping of related data can offer key inputs for water security to cross-sector impacts, helping address questions such as:

- What's the role of integrated or "nexus" solutions that include revolutionary data integration and transparency?
- How will responses to different hazards—e.g., extreme heat, drought—enable new resilient infrastructures and institutions that bridge public-private-entrepreneurial innovation?
- What if tomorrow's systems and services had "urban nexus solutions" (e.g., integrated resource efficiency, circular economy, renewable energy desalination, water reuse for food and energy production, integrated infrastructure and institutional (re)development, and modernization to address aging, centralized, legacy systems that are functioning poorly)?
- What are the best urban water, energy, and food pathways that efficiently advance water security, with radical transparency, offering "leapfrog" improvements to quality of life for all, complemented by decoupling of economic prosperity from environmental impacts?

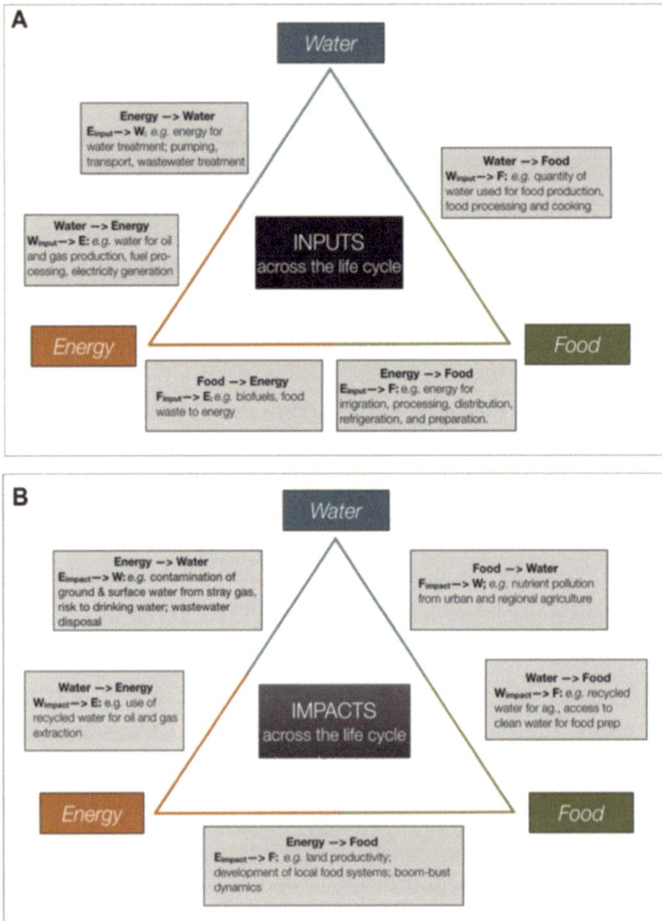

Figure 1. Illustration of initial pairwise relations in the FEW nexus framework for a proposed extension of an urban nexus maturity model, to inform and help develop nexus indicators at the urban-regional scale, considering (a) inputs to food, energy, and water systems as well as (b) impacts on each of these systems from the other two. Source: Ahamed et al. [5].

4. New urban nexus concepts for resilience and abundance

Water and energy are national and global priorities for security, economic prosperity, and human well-being. The term "coupling" has been applied to communities and nations that have effectively enabled new synergies between water and energy industries, technologies, infrastructure, or policy trajectories that maximize economic prosperity (or productivity) while enhancing resource or service sustainability (or resilience), respectively. Similarly, "decoupling" often refers to the positive outcome of increasing economic prosperity while reducing environmental impacts and unsustainable resource use. The term "leapfrogging" has also been applied to communities and nations that have adopted new forms of advanced infrastructure, technology, and cooperation, so as to reduce risk among competing uses through new insights. By bypassing traditional progressions, e.g., from no phones to cell phones—bypassing landlines altogether - the overallocation of resources or increasing competition for scarce freshwater sources could be bypassed through integrated approaches to water, energy, and agricultural productivity. For the first time in history, transitions and transformations (perhaps even "leapfrogging" opportunities) exist for energy-efficient water services and water and energy-efficient agricultural productivity across rural, suburban, and urban environments of the USA. Such innovation can be further catalyzed within a context of prizes and critical water issues for global geographies, with a focus on market-based solutions that enable national competition through improved energy and water services for smart, healthy, and resilient communities.

While aging and deficient infrastructure systems for water, wastewater, transportation, and energy have become the norm in the developed economies, humanity is experiencing new frontiers in digital technologies, globalization, and water-energy connectivity. The trends of urbanization [6], in particular, have many countries and their communities in emerging markets asking how and why water technology and infrastructure modernization could help to increase competitiveness in attracting talent and businesses, enabling new "leapfrogs," and decoupling economic prosperity from environmental damage. Setting triple bottom-line goals—social, economic, and environmental—while harnessing new technologies, training-entrepreneurship-innovation roadmaps, and multi-sector approaches that meet rising demands for services and while not being hampered by legacy, aging, or deficient systems—may offer cities and nations a competitive advantage within a global economy.

To keep pace with growth, it is estimated that by 2030—just 12 years from now—the world may have to produce 40% more clean water, 35% more food, and 50% more energy (UN, World Water Development Report, 2014). Moreover, these basic resources are mutually interdependent—to produce more of any one of the resources requires more of either one or both of the other resources. Increasingly, these resources will have to be made more accessible to locations where the majority of populations will reside in years ahead or to areas increasingly exposed to risk and vulnerability.

A fundamental step to address these challenges is to improve the efficiency of water and energy use. Intrinsically linked, wasteful operating practices in both energy and water systems increase the scarcity and cost of these resources. Both energy and water systems require focused efforts to optimize resource use. The cost and energy savings associated with improving water/wastewater system efficiency through variable frequency drives, fixing pipeline leaks and breaks, and tightening building envelopes to reduce energy loss are well understood. The key challenge to the adoption of these efficiency measures is funding. Funding gaps by power and water/wastewater utilities create significant obstacles to realizing these simple system improvements.

The following sections offer new city-related concepts and exhibits into the people and communities investing in transformations in urban systems related to water, energy, food-related infrastructure, governance, and new behaviors. These exhibits also help respond to the identified 20 top challenges facing the urban water sector (as identified in an AWWA report: top 20 challenges facing water sector (2018 survey results) in https://www.awwa.org/Portals/0/files/resources/water%20utility%20management/sotwi/2018_SOTWI_Report_Final_v3.pdf.

4.1. Exhibit A. Designing integrated water-energy-food technology hubs

Digital solutions. John Deere is now using sensors in several of its products to increase farm productivity (http://www.bigdata-startups.com/BigData-startup/john-deere-revolutionizing-farming-big-data/). John Deere uses sensors added to its latest equipment lines to help farmers manage their fleets and to decrease tractor downtime while also saving on fuel. Sensor information is combined with historical and real-time data on weather prediction, soil conditions, crop features, and many other data sets. These tools aim to increase the productivity and efficiency of crops, resulting in higher production and revenue.

RENEWW Zones. By 2030, impoverished peri-urban areas are expected to double in size to almost two billion people. Rapid growth in resource demands due to population growth is already outpacing many governments' ability to extend basic services to slums and informal settlements, while centralized legacy water and sewer infrastructure systems are breaking down. How will communities that cannot meet their populations' needs for sustainable water, food, and energy now be able to meet them in future?

In response to this challenge, the USA is catalyzing a Renewable Energy, Nutrition, Environment, and Water and Waste resource recovery initiative (RENEWW). Innovation zones co-design partnerships between civil society, businesses, academia, and grassroots organizations to build capacity and harness innovations at the nexus of food, energy, water, and other systems that support and foster inclusive, smart, sustainable, healthy, and resilient communities that leverage the best of US innovation. To stimulate the development of game-changing solutions, RENEWW is launching a highly leveraged, incentivized community prosperity prize competition that pushes the limits of what's possible, captures the world's imagination, spurs new thinking, and accelerates change through the creation of RENEWW Zones.

RENEWW Zones are decentralized, closed-loop models of spatial planning and peri-urban service provision that replace fossil energy with renewables; derive new water, biogas, and fertilizer from wastewater; and produce food and biofuel with the recycled inputs, all cogenerated at near net-zero waste. Each RENEWW Zone would offer a green space for community recreation, recycling and sanitation services, as well as a place to purchase fresh food, recycled goods, biofuels, and safe drinking water, all within walking or cycling distance. Ideally, well-planned RENEWW Zones placed at the outer edge of existing informal settlements would provide a basis for adjacent planned urban extensions. RENEWW Zone business models would create local employment, reinvest profits to support operational costs, and engender new public and private financing. Profitable Zones would scale through replication as local private investors realize the potential profit in serving society's bottom billion.

4.2. Exhibit B. Off-grid "circular economy" services—affordable, accessible, and reliable

Zero Mass Water (https://www.linkedin.com/company/zero-mass-water) has built a residential solar air moisture capture system that can provide safe drinking water for a family of four. The system has been deployed in Ecuador, Jordan, Mexico, and the USA. Essentially, off-grid, safe drinking water is powered by solar energy.

The concept of "circular abundance" has been introduced recently for brainstorming new water and energy-related challenges and prize competitions, and the concept focuses on capitalizing on the innovative integration of energy- and water-efficient businesses, technologies, or other resources that flow from one to another in a synergistic, sustainable manner. It envisions a closed-loop model of responsible conservation and economic development that replaces fossil fuels with renewable sources; derives new water from wastewater, rain, and agricultural water; and produces food with recycled energy and water while creating near net-zero waste.

This nexus model aggregates and co-locates solutions—such as vertical farming; blue, green, and solar roofs; and waste-to-energy technologies—so as to allow the movement and utilization of resources easily from one production facility to the next. For example, using waste from a fish farm and converting it to energy to run the facility and then incorporating other technologies like efficient water filtration, water reuse, LED grow lights in vertical green houses, and solar panels can make the location ecologically and financially self-sustaining. Sewage water can be transformed into drinking water, electricity, biogas, and ash. And polluted water can be processed to extract fertilizer, industrial chemicals, and metals for reuse/resale. And by co-locating these production processes, we will maximize the efficiency of each individual process, minimize the amount of raw resource inputs (such as fuel, water, and land) required, and eliminate waste. Creating a system of "circular abundance" requires holistic water-food-energy planning—but in large-scale disconnected systems, the technical and social complexity of the challenge generally overwhelms even the most qualified urban sustainability planners.

4.3. Exhibit C. Conservation synergy and blended finance

Conservation synergy and blended finance. In 2008, the investor-owned, California-based utility PG&E, along with several water agencies in California, offered a rebate program for high-efficiency clothes washers. The rebate in 2013 ranged from $100 to $125—this includes a $50 rebate from PG&E and a variable rebate from $50 to $75 from the water utility. PG&E has seen a 63 percent increase in customer participation since the water utilities joined the program and the water utilities have seen a 30 percent increase in their customer participation. The program has since expanded to 41 municipal, regional, and private water utilities.

We are also seeing a movement toward "blended finance" which, as the name implies, brings together diverse sources of capital to fund much needed investment in infrastructure. Two of the key actors in this movement is OECD and University of Oxford (OECD, 2017. Blended finance: mobilizing resources for sustainable development and climate action in developing countries and OECD-WWC-Netherlands Roundtable on Financing Water Second meeting 13 September 2017, Tel Aviv Session 4. Background paper The potential for public, purposed, development and hybrid finance to bridge the water infrastructure gap Alex Money, University of Oxford).

There is also great opportunity for the private sector to drive competition and innovation to help water and wastewater utilities adopt efficiency measures. Energy savings performance contracts are established mechanisms used in a variety of sectors to implement building energy improvements; however, these mechanisms are scarcely used in the water/wastewater arena. A challenge specifically focused on fostering private investment in the water/wastewater sector would reduce both energy use and costs. Through innovative finance models, the private sector would foster a measurable and significant impact on energy use in the water/wastewater sector while also providing capital to help these utilities with the investments needed to address the growing challenge of aging infrastructure. This effort would help to build capacity within the water/wastewater sector on energy use and conservation opportunities while overcoming the perceived obstacles of private sector investment in their systems. It would further incentivize private sector investors to evaluate the business propositions of water/wastewater utilities, identifying new opportunities for US business growth. The energy conservation measures addressed through this financing would be measurable in both cost and energy savings. The end result would be robust water/wastewater utilities that have optimized system performance to conserve energy and cost, freeing up operating budgets to invest in infrastructure repair and replacement programs.

4.4. Exhibit D. Urban, behavioral, and decision science on public-private partnerships

While humans historically planned for choices related to one-way flows of energy, water, and information (e.g., from a TV/radio, water/wastewater treatment and power plants), new energy and water technologies and their integrated services affording two-way and multi-directional information, energy, and water flows (and feedback for increased efficiency and economic opportunity) have the potential to become essentially ubiquitous in the decades to come. However, the opportunities to provide targeted services that optimize human prosperity, energy, and water benefits remain vastly underutilized and under-imagined.

Moreover, many institutions still appear slow to recognize and respond to the fact that water and energy systems can now behave more nimbly and adapt in real time as a result of recent disruptive changes in technologies and services. A prize is needed to address growing service and industry demands (e.g., for water management in hydraulic fracturing, (waste)water reuse, water for energy/agriculture, asset management, energy in water, and wastewater systems).

By creating a "race" to address the rapid pace of techno-economic change, increased connectivity, and transition opportunities toward improved services, prizes can improve the cost efficiency of water and energy services, and communities could continue to move toward circular economies with resource-efficient systems (including considerations that bring together energy, water, agricultural productivity, land, materials, waste, etc.). These technology and infrastructure service disruptions, if guided by prize challenges that enable higher public benefit, can provide for rapid expansion in choices, new finance revenues, and potential "leapfrogs" in enabling new businesses and industries that improve resident and business siting for co-locating near better services (faster, cheaper, safer, more reliable, cleaner, higher quality). Hybrid decentralized-centralized systems are increasingly viewed as key, and new data and analyses that evaluate the "decoupling," "leapfrogging," and "competitiveness" potential of new, integrated, market-based approaches will be needed.

A growing area of interest is at the intersection of water and energy system operations. These systems are often disconnected and operated independently. However, when considered in tandem, system integration may enable new opportunities to revolutionize the current paradigm of thinking. Energy use in the water/wastewater sector is poised to grow in order to meet the demands of population growth, deteriorating water quality, and increasingly stringent water regulations. Widespread deployment of variable electricity generation (e.g., wind and solar) is also placing a premium on power system flexibility. The extent to which drinking water and wastewater systems are powered by electricity and can be operated flexibly due to their inherent storage capacity and deferrable loads highlights the growing importance of the relationship between these two critical infrastructure systems and the need for integrated, resilient energy and water systems that perform reliably under normal conditions and are prepared for and can recover from disruptive events.

The power grid is continuously balancing supply, demand, and power quality requirements. Ancillary services are services provided to the grid that help match supply to demand and maintain power quality. Controlling or changing loads to support the grid or "demand response" is a strategy used by power system operators to balance supply and demand. Water/wastewater systems are beginning to explore participation in demand response programs, but they are still not widely adopted.

Multinationals and nongovernmental organizations (*NGOs*). The 2030 Water Resources Group (http://www.2030wrg.org/) released an online database of case studies to address water scarcity risks. It is designed to facilitate adoption of leading practices to cover a wide range of common scarcity challenges as well as proven solutions. The group offers for free download the full catalog of in-depth solutions (http://www.waterscarcitysolutions.org/).

These interdisciplinary and integrated responses have breakthrough and "leapfrog" potential for improving quality of life and the sustainability and resilience of communities in the context of water.

4.5. Bridging the rural-urban divide: harnessing emerging technologies and services, strategic planning, policy, behavior change, and finance

Today, characterization of the critical urban-to-rural water-energy-food sustainability considerations is often lacking, despite increased realization that the drivers of urbanization and city demands also have critical rural impacts—especially under quickly changing economic/environmental dynamics.

Going forward, can we predict—using data and multiple metrics—where and when rural and urban environments will see vulnerable communities become hotspots of vulnerability and/or seedbeds of innovation? For example, this could include demand/supply/ecological quantity/quality—and identifying where gaps are growing and where climate may add risk—and potentially bring new ethical dilemmas and painful choices. By first taking a look at some key consequences from already unfolding urban water crises, it may be possible to unpack several key messages/hypotheses for further rapid experimentation and evaluation of effectiveness.

This can include city strategies and objective assessments that include the ability to:

- Define and measure innovation, reinvention, and maturity of responses.
- Identify sweet spots of service users, designers/operators, and cross-scale policy actors (including public, civil society, business/firms) and their roles in rural to urban settings.
- Proactively move responses at the speed of need by creating continuous dialog and exchange, communities of practice, and feedback loops.
- Predict, mitigate, and finance immediate responses to long-term strategy (with evaluation of past failures, ethical dilemmas, and policy processes) for enabling secure and resilient cities.
- Expose future interurban risk/vulnerability hotspots and resilience strategies (as maps).

5. A path forward: urban maturity framework

How do we move past business as usual, and what is the road map or response framework for cities? Can cities and nations achieve competitive advantage through integrated sustainability and resilience strategies for transformation toward secure, affordable, reliable water systems and services? Themes emerging in this chapter and across the water industry pointing to new, integrated opportunities increasingly include:

- Personalized and valued services
- Shared/circular economy/closed-loop services
- Data/transparency for agile development
- Multidisciplinary approaches to breaking down boundaries and enabling new finance
- Diversification and decentralization with limited hierarchy enabling new accessibility
- Advocating and measuring for "excellence" using creative, entrepreneurial approaches

While there are several answers to move cities to a thriving, resilient state with equitable access to safe drinking water, there is no silver bullet. Clear examples exist on two sides of the same coin (as to human behavior/decisions): 'necessity as the mother of innovation' and 'if it ain't broke, do not fix it, both of which are business as usual responses. In essence, how do we avoid business as usual failures in Cape Town, South Africa, that have occurred also across the many other cities?

Certainly, what is needed are better data and actionable information. However, we also need a process to use actionable information to inform public policy innovation to accelerate scaling of innovative technologies, funding and financing strategies, and partnerships and business models. To a significant extent, we need to mobilize all stakeholders, including but not limited to the public sector, to ensure water to drive economic vitality (Averting the Next Cape Town).

"Ultimately, water scarcity is a challenge for society as a whole, which needs to gather the will to develop a strategic water plan to avert the otherwise inevitable 'Day Zeroes' to come."

What has to change? Here are six steps viewed as essential yet only a beginning:

1. *Engagement*: If the public is unaware and/or does not care, we are destined to see more Cape Towns. It is the responsibility of the public sector to provide access to safe drinking water (SDG 6) but also the responsibility of the public to support such efforts.
2. *Innovation*: We need a broader view of innovation beyond technology to include business models, financing/funding, public policy, and partnerships.
3. *Scale*: Innovative solutions need to be scaled which requires adequate and sustained funding from consumers, government, and the private sector.
4. *Pricing*: Access to water is not free; it is now a costly necessity. We need to treat it as such by recognizing that access to water requires investment to protect and provision equitably.
5. *Urgency*: Voluntary approaches will not address the water crisis. We need regulatory action that fosters innovation and enforces conservation and does it quickly.
6. *Honesty and transparency*: The public sector must acknowledge a new normal, and investments are required to ensure access to water. We no longer can blame the weather. As long as we continue to refer to Cape Town as a natural (rather than man-made) disaster, hope will be the best water strategy we can muster.

Given a shared responsibility to manage our scarce fresh water supplies, self-interest demands action. All individuals are ultimately accountable for this shared resource, both to each other and to future generations counting on our decisions to get this right.

What is the road map for cities? An urban water strategy maturity model is a useful framework to guide cities in developing an actionable strategy to avoid a "Day Zero." Below offers a timeline observing the evolving decision systems that are moving from reactive emergency/crisis response modes, voluntary programs, and inadequate data systems to proactively accepting responsibility for informed responses to increasing frequency and intensity of extremes (**Figure 2**).

In this maturity model, a road map can move cities from either no strategy or a severely outdated strategy to one where abundance with regard to water is created. Abundance is feasible but only if we acknowledge our current reality and leverage all stakeholders to apply the appropriate technologies and strategies available in the twenty-first century.

A critical aspect of moving along the maturity model is ensuring access to granular data and actionable information. For example, values for water/wastewater sector energy use are either high-level estimates (e.g., US water pumping, treatment, and distribution in 2015 required 34.65 billion kWh) or the energy use of specific processes (e.g., ultraviolet disinfection requires 255.5 kWh/million gallons treated). These data are insufficient for researching energy-water utility interactions on a city or national scale as energy use is a function of local water quality, regulations, technologies used, populations served, and geographical layout of systems.

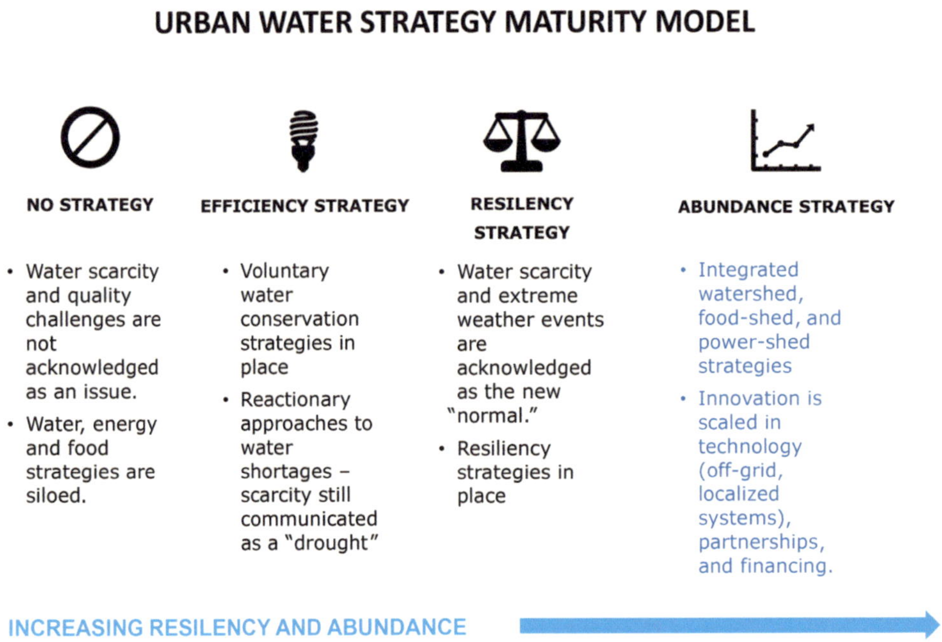

Figure 2. Urban response maturity model.

A solution to this critical challenge is to create an industry-validated energy dataset for water and wastewater utilities, and a water dataset for municipal, energy, and agricultural production activities helping to inform and quantify demand response opportunities for water/wastewater utilities and cities and provide new insights into the complex flows between electricity, agriculture, and water/wastewater utilities in the USA. Through a competitive prize challenge, entities could submit innovative approaches to collect and maintain this data, make it publicly available, and apply it to quantify, at a national level, the opportunities that exist for water/wastewater utilities to provide grid services.

This type of concept could define and enrich the understanding of how energy is used in the water and wastewater sectors or vice versa, as well as the potential for utility services that these systems can provide, helping to bridge the knowledge gap between water and energy systems and advancing maturity in this area. Identifying potential energy-water value propositions to water/wastewater utilities and the electric utilities that serve them is key. Further, such work can foster future dialog, technology, and policy solutions to complex challenges of integrating water-energy-"X" systems.

6. Conclusions: partnerships to inform smart and resilient systems

Additional research and practitioner efforts are needed to develop a network and community of research and practice that brings together interdisciplinary innovations from universities and national labs to the private sector who will coordinate their work, with focus on assessment of and enabling decision-making. This can include foci of water yet also broader emerging technologies coupled with new insights into human behaviors and decision processes within contexts of urban crises. Urban dimensions of mitigating risks to and increasing productivity of water infrastructure services are now more critical than ever. Further, road maps could aim to explore how, why, and where connected, automated, decentralized, and 100% renewable energy-driven water treatment systems and the critical services they enable can help achieve smart, sustainable, healthy, and resilient cities.

Future research and analyses could focus on at least three urban sites moving from painful decisions to abundant choices and increased resilience that has returns on investment (e.g., from new revenues enabled via system integration). In-depth case studies over several years may also be needed in order to inform effectiveness of responses to acute challenges as well as the longer-term design, operation, and use of systems/services in urbanizing to aging/legacy infrastructure-dependent cities.

Existing and new data streams coupled with visualization tools as means to provide interdisciplinary teams with evidence for water security, water-efficient energy, and agricultural systems to energy-efficient water delivery systems and infrastructure could all prove valuable.

Maturity models and frameworks that operationalize the extent to which urban areas are harnessing emerging data technology-human behavior-decision processes that can help enable transformations is a key message of this chapter, building on the characterization of the extent to which current challenges exist, motivating conditions for (revolutionary and disruptive) change to future prospects of sustainable trajectories toward the security and resilience of people, infrastructures, and resources.

New integrated road maps have the potential to substantially increase water and energy productivity, affordability, and resilience of urban (and trans-boundary regional) infrastructure. These infrastructures not only refer to the maturity of the technological and built environments yet also the diverse engineered-natural-social-cyber systems that provide water, energy, goods, and information services to more than 50% of the population living in cities today.

With rapidly increasing populations, projected resource scarcities, and vulnerability to disasters, smart and resilient cities will require new, high-performing, cost-effective infrastructures for future water (and energy) systems. One overarching question to be addressed moving forward might be: 'What are the interconnections of sectors, disciplines, and decision-making domains that must be explored to design sustainable, smart, resilient, and modern urban water and infrastructure systems of the future?'

Author details

Will Sarni[1]* and Josh Sperling[2]

*Address all correspondence to: will@waterfoundry.com

1 Water Foundry, Denver, Colorado, United States

2 National Renewable Energy Lab, Denver, Colorado, United States

References

[1] Arup. The Future of Urban Water: Sao Paulo. 2016

[2] eNCA. Bangalore Faces Man-Made Water Crisis. 2018. https://www.enca.com/world/bangalore-faces-man-made-water-crisis [Accessed: July 2018]

[3] Rashad J. The world's longest river is in trouble. The Washington Post. 2018. https://www.washingtonpost.com/news/theworldpost/wp/2018/03/22/egypt/?utm_term=.826b245a97c9 [Accessed: July 2018]

[4] Tellman B, Bausch JC, Eakin H, Anderies JM, Mazari-Hiriart M, Manuel-Navarrete D, et al. Adaptive pathways and coupled infrastructure: Seven centuries of adaptation to water risk and the production of vulnerability in Mexico City. Ecology and Society. 2018;**23**(1):1. DOI: 10.5751/ES-09712-230101

[5] Ahamed S, Sperling J, Galford G, Stephens J, Arent D. The food-energy-water nexus, regional sustainability, and hydraulic fracturing: An integrated assessment of the Denver region. Case Studies in the Environment. 2019;**2019**:1-21. DOI: 10.1525/cse.2018.001735

[6] McDonald RI, Weber K, Padowski J, Flörke M, Schneider C, Green PA, et al. Water on an urban planet: Urbanization and the reach of urban water infrastructure. Global Environmental Change. 2014;**27**:96-105 https://www.sciencedirect.com/science/article/pii/S0959378014000880 [Accessed: July 2018]

Chapter 4

Water, Ecosystem Dynamics and Human Livelihoods in the Okavango River Basin (ORB): Competing Needs or Balanced Use? A Review

Ketlhatlogile Mosepele, Wame L. Hambira,
Goemeone E.J. Mogomotsi, Patricia K. Mogomotsi,
Oliver Moses, Masego Dhliwayo,
Anastacia Makati and Bernice Setomba

Additional information is available at the end of the chapter

http://dx.doi.org/10.5772/intechopen.80554

Abstract

Freshwater is essential to life, and its availability poses a significant challenge to developmental needs and environmental sustainability globally. Due to increasing populations, global water requirements have increased in the twentieth century, and the trend is similar in the Okavango River Basin (ORB). With a total annual flow of 11 km^3, the ORB is characterised by a flood pulse regime that drives and supports a diverse eco-sociological system. The Okavango River is a potential water source for the development of the semi-arid nation states of Botswana and Namibia. Therefore, there is a need to ensure that the water resource of this system is managed effectively to ensure water sustainability in the basin. Current water demand in the basin is less than 1% of the current total discharge, while projected demand over the next 10 years also falls below the total discharge. Moreover, the ORB is characterised by multi-functional use, where riparian communities have adapted to change hydrological conditions. While the ORB is relatively pristine, there are potential threats in this system, which can affect its water resources. We conclude that there is a need for a harmonised legislative framework in the basin to ensure that the ethos of water sustainability is maintained.

Keywords: water scarcity, water management, water governance, transboundary water resources management

1. Introduction

There are 214 transboundary river basins in the world [1], which cover large parts of Arica, Europe, Asia and the Middle East, and were inhabited by 58% of the world population in 2010 [2]. This makes transboundary river basins critical foci of human livelihoods. They are a source of freshwater, which is essential for life [3, 4]. The world's land and water resources are finite and under pressure from population growth [5], climate change and pollution [6]. Water resources management becomes more challenging when water resources straddle international boundaries [2], such as in the Okavango River Basin (ORB). Floodplains have always played a key role in human livelihoods [7] and also have the ORB [8]. According to Junk [7], hydrology is the key "environmental forcing factor in wetlands"; therefore, maintenance of natural flow conditions is critical for the environmental sustainability of the ORB. Hence, management of the ORB is critical towards sustenance and maintenance of human life among the basin states. However, humans have always withdrawn freshwater from sources such as rivers, wetlands and lakes for various needs such as agriculture, energy and industrial activities, at the exclusion of ecosystem needs [3]. Therefore, ensuring water sustainability in the ORB will require finding a balance between competing human needs and ecosystem functioning.

The waters of the Okavango River are a potential source of development for the semi-arid nation states of Namibia and Botswana [8]. Therefore, water is both a key and a limited resource in the ORB [9]. Because of its uniqueness, national socio-economic development polices have added pressure on the water resources of this river system [10] in their endeavour to uplift the socio-economic status of the basin's impoverished communities. However, water sustainability becomes more urgent and acute in transboundary river systems because of the diverse hydro-political and socio-economic drivers that exist. Munia et al. [2] highlight that transboundary rivers create hydrological, social and economic interdependencies among societies, which make transboundary water resources management challenging.

Some of the key transboundary management areas of concern identified by OKACOM [10] in the ORB are variation and reduction in hydrological flow, changes in sediment dynamics, changes in water quality and changes in the abundance and distribution of biota. These issues are driven primarily by population dynamics, land-use change, poverty and climate change. According to OKACOM [10], the major transboundary concern is that increasing populations in the basin will result in increased demand for food crops with subsequent pressure on land, which will invariably result in changes in water quality. Undoubtedly, well-managed water resources can be a significant driver of growth with benefits for human livelihoods and ecosystem functioning [11]. Therefore, the main goal of this chapter is to contribute knowledge that will contribute towards a water resources framework for the ORB with the aim of achieving water sustainability.

2. Description of the study area

The ORB is located in central Southern Africa (**Figure 1**) and covers a broad climatic gradient from a high rainfall zone in the Angolan highlands, through a semi-arid Namibia and ends in the semi-arid Northern Botswana [12]. It covers a total surface area of between 20,000 Km^2 [13]

and 690,000 km², with a population of approximately 900,000 [5, 10, 14]. According to Barnes et al. [15], the population in the basin is predominantly rural and remote and has higher population growth rates than national averages. The basin's population is expected to grow to 1.28 million people by 2025 with 62% of the population living in Angola, 22% in Namibia and 16% in Botswana [10]. Assuming a medium variant growth, the river basin population will increase to 5.1 million people in Angola but will plateau in Botswana and Namibia by 2050 [16]. This is expected to increase pressure on the ORB, especially in Angola [9].

According to Wolski et al. [12], the ORB is composed of an upper part (in Angola) characterised by a typical river catchment and a lower terminal part (in Botswana) consisting of the Okavango Delta and terminal rivers, where the "water's ultimate sink is evaporation to the atmosphere". The ORB has four catchment areas, which the Angolan headwaters (in Angola), the middle reaches (in Namibia) and the Botswana portion, which consists of the panhandle and the delta ([10, 17], Figure 1). This system is drained by the Okavango River, which is one of the largest rivers in Africa [13]. Because of its location in an arid area, the Okavango River is a major source of water in the region [13].

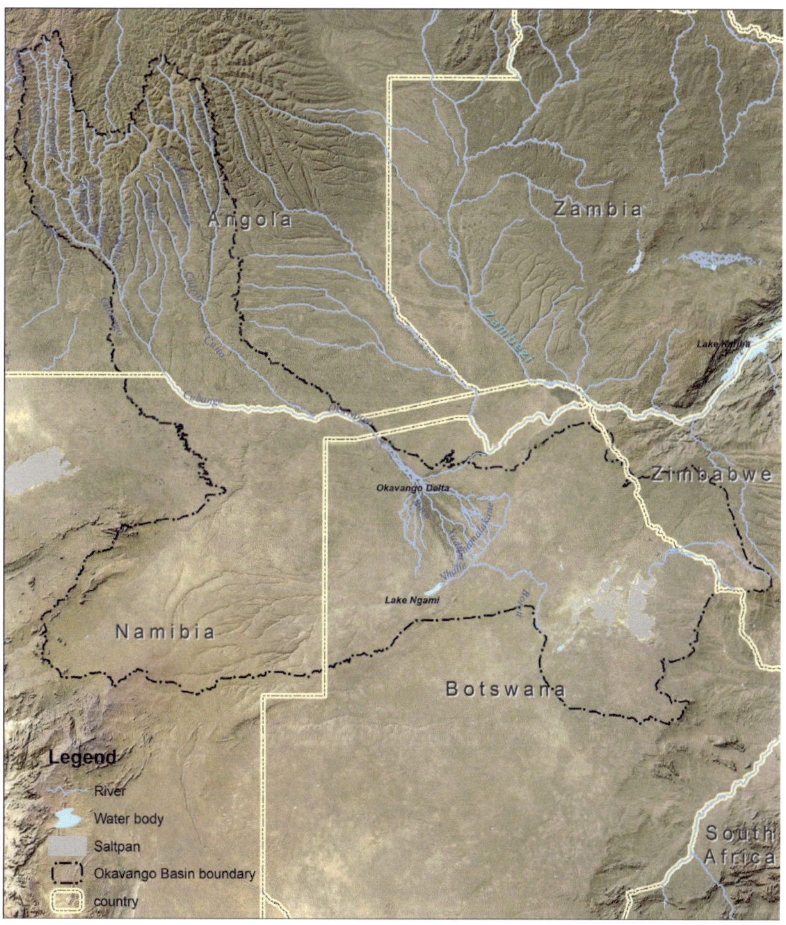

Figure 1. The Okavango River Basin (ORB).

Total annual water inflow in the ORB, characterised by an annual flood pulse, is approximately 10,900 Mm^3 [18], while that of the lower basin is 9600 Mm^3 [10, 18, 19] and drought flow is 3120 Mm^3 [18]. Peak flow of this flood pulse from the Bie Plateau in Angola reaches the delta in Botswana between February and April [19, 20], which coincides with the end of the rainy season in Botswana [20]. All the water flow in the system is generated upstream of the confluences of the Cubango and Cuatir Rivers in the west and the Cuito and Longa Rivers in the east [10], with a combined area of 38,700 km^2, contributed 5185 Mm^3/annum of water into the system, which is 48% of the total water volume in the ORB [10]. In fact, the Cuito River is the most important for downstream flows into Namibia and Botswana [18], hence any changes to water flow in this river system will have a significant impact on discharge into these two countries. Overall, the Angolan catchment is approximately 13,000 km^2 and contributes about 95% of the water inflow into the system [14]. There is multi-decadal-scale variability in rainfall, temperature and discharge characterised by 30-year non-overlapping periods in the basin [21].

According to Pröpper et al. [22], woodlands on Kalahari sands are the major land cover type in the ORB, followed closely by Miombo forests. Other land cover classes in the ORB are thorn-bush savanna and shrub and grasslands, while wetlands constitute 7.3% of the land cover in the basin. Furthermore, SAREP [23] observed that Miombo woodlands dominate the upper catchment of the basin, which graduates into deciduous woodlands in the middle reaches (Namibia), which then turns into mixed acacia and mophane woodland in the lower reaches (delta). The delta, a key biodiversity hotspot in the basin [20, 24], hosts approximately 1300 plant, 444 bird, 122 mammal, 64 reptile, 33 amphibian and 71 fish species [19].

The ORB is one of the least developed basins in Africa [4, 10, 13] and supports predominately rural communities [10] with a relatively low population density [4]. It is relatively pristine, possibly due to a strong conservation ethic in Botswana, which is focused on tourism, poor agricultural soils and a civil war that ravaged the catchment area (Angola) from 1970 until 2000 [10, 12]. However, the end of the civil war has resulted in a rapid population build up in the catchment [12], which might result in development pressures on the basin in the future. Poverty rates among communities living within the ORB are much higher than the national average among the three countries [5]. Generally, poor people depend more on natural resources as social safety nets than communities in urban areas, which can invariably lead to environmental degradation [15].

3. Theoretical framework

This study used primary and secondary data, which was then integrated with literature review to produce a state-of-the-art analysis of water sustainability in the Okavango River Basin. The transboundary water assessment programme [25] framework was used in this study to assess water sustainability in the ORB. There are five core elements of the framework, which include water quantity, water quality, ecosystems, governance and socio-economics. These core elements have 15 indicators associated with them as summarised in **Table 1**. We added a 16th indicator, which reflects the importance of climate change as a major factor that needs to be incorporated into water resources management.

3.1. Water quantity

The global population is growing steadily, and this will result in a corresponding increase in water demand for food production, especially from wetlands [26]. Invariably, this may result in water scarcity, which is defined by Matlock [27] as a function of available water resources for the human population, where the *Falkenmark* indicator is a widely used indicator of water stress. The different *Falkenmark* categories are shown in **Table 2** and range from no stress to absolute scarcity.

African growth is expected to be driven largely by primary and secondary (economic) sectors, which are heavily reliant on water, with irrigation as a key food production activity [6]. However, agricultural water demand (agricultural water stress) in the basin includes livestock water and irrigation [5]. Indicators of agriculture-induced water stress in some global basins include "closed basins, reduction of groundwater resources, loss of wetlands and habitat fragmentation" [26]. Therefore, there is a need to ensure that agricultural water demands are balanced against ecosystem water needs, which are often defined as environmental flows. According to Forslund et al. [28], these refer to the quantity, quality and timing of flows that are needed to sustain ecosystems. These flows are partitioned between ecosystem needs and other key users such as agriculture, power generation, domestic use and industry [29].

Core element	Indicator
Water quantity	1. Environmental water stress
	2. Human water stress
	3. Agriculture water stress
Water quality	4. Nutrient pollution
	5. Wastewater pollution
Ecosystems	6. Wetland dis-connectivity
	7. Ecosystem impacts from dams
	8. Threats to fish
	9. Extinction risk
Governance	10. Legal framework
	11. Hydro-political tension
	12. Enabling environment
Socio-economics	13. Economic dependence on water resources
	14. Societal wellbeing
	15. Exposure to floods and droughts
Cross-cutting	16. Exposure to climate change

This is adapted from UNEP-DHI and UNEP [25].

Table 1. Summary of the five core elements and associated indicators of the transboundary waters assessment programme (TWAP) that underpins the theoretical framework for this study.

Index (m^3 per capita)	Category
>1700	No stress
1000–1700	Stress
500–1000	Scarcity
<500	Absolute scarcity

Table 2. Summary of *Falkenmark* categories used in the assessment of water scarcity in the ORB.

3.2. Water quality

River catchment degradation is a key issue of concern in contemporary river basin management in tropical systems. This degradation is driven by increasing population pressures, which place a heavy burden on natural resources [30]. Therefore, water quality management in river systems is critical towards controlling river pollution in which land use is a critical component of water quality in river basins [31]. The key land-use types that affect water quality in river basins are urban and agricultural activities, whose key indicators are elevated concentrations of bacteria, pesticides and nutrients [32]. Failure by governments to preserve water quality of surface waters, especially in river basins, may enhance fragility of communities [33].

3.3. Ecosystems

Wetlands are a key source of goods and services but are threatened by unsustainable use, both within wetlands and also in the upstream catchments [34]. They are biodiversity hotspots and foci of high biological production at the water-land interface [35] and are critical for biodiversity conservation [36]. Some key wetlands' resources include freshwater for human livelihoods, fish and flood protection [37]. Wetlands play an important role in river basins, while human activity within those river basins can have a negative impact on wetlands [36] usually driven by population growth [38], land-use change and climate change [10]. Because wetlands provide benefits to basins [39], there is a need to integrate their management into basin management [36] to ensure that this beneficial relationship is sustained.

3.4. Governance

Proportionally, water crisis is the top most important issue among the top five (i.e., water crises, failure of climate change mitigation and adaptation, extreme weather events, food crises and profound social instability) global risks of the highest concern for the next 10 years [33]. Therefore, governance of water resources, especially in river basins, is critical towards achieving water security. Governance considers multi-level participation beyond the state and includes the private sector, civil society and society in general [40]. Orme et al. [41] propose the Good Transboundary Water Governance Matrix (**Figure 2**) as a key tool in water governance of shared water courses. This matrix incorporates the sustainable development goals (SDGs) and is also based on the United Nations Watercourses Convention [41]. According to Sadoff et al. [33], the fragility of basin communities can be heightened by failure of governments to preserve transboundary water resources, usually underpinned by weak institutions. Therefore, good water governance should include shared water courses [6], which will ameliorate any hydro-political tensions among basin states.

Sustainable transboundary water management is anchored upon substantive and procedural criteria ([41], Figure 2). The substantive criteria are circumscribed by three legal obligations. The "equitable and reasonable utilisation" criterion is the cornerstone of international water law and is the anchor for transboundary water governance, while the "duty not to cause significant harm" criterion refers to limiting pollution or over-exploitation, which might have a negative impact on the environment. The "protection and conservation of ecosystems" criterion is self-explanatory. The procedural criteria are anchored on four key obligations. The "notification and information exchange" criterion is implemented when parties notify others of planned developments that might negatively affect other users. The "environmental impact" criterion refers to a process of making informed developmental decisions based on a thorough analysis of anticipated environmental impacts. This criterion also provides a platform for community participation, which is also explicitly stated in the "public participation" criterion, and refers to the obligation to consult the public. When the "access to justice" criterion is upheld, then "information exchange and public participation rely on enforcement and review mechanisms to ensure efficacy and equity" [41]. Furthermore, this framework can create an enabling legislative environment that can reduce/minimise hydro-political tensions between the basin states, through by creating an enabling environment.

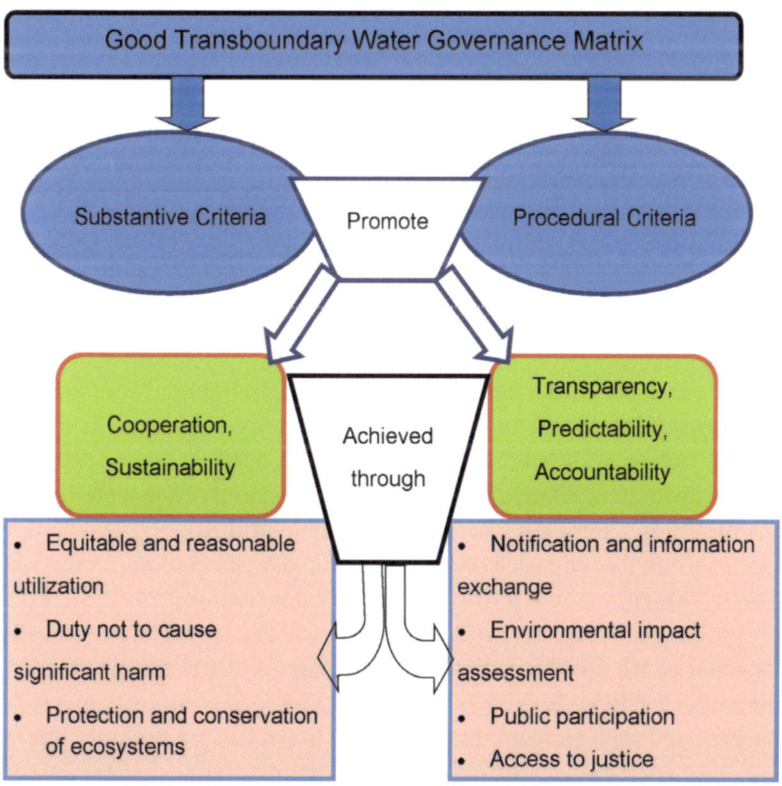

Figure 2. The Good Transboundary Water Governance Matrix reproduced from Orme et al. [41].

3.5. Socio-economics

The livelihoods of the majority of rural African populations are intertwined with water [6]. This is because water is the basic foundation of human livelihoods [26], and water scarcity can have a profoundly negative impact on economic growth [42]. According to UNEP-DHI and UNEP [43], the key components of the coupled human-environment system used in the assessment of river basins are "economic dependence on water resources, societal wellbeing and exposure to climate-related natural hazards". This suggests, therefore, that the observation of water being not only a key resource but also a limited resource in the ORB [9] makes it vulnerable to future population growth in the basin.

3.6. Economic analysis

Water resources management initiatives at macro and micro levels are faced with positive transaction costs [44, 45]. The economic considerations of allocation, efficiency, equity, production and pollution have significant influence on water-related decision-making, water resources management and water policy formulation. However, because it assumes a world of zero transaction cost, neoclassical economics "is incapable of handling many of the problems to which it purports to give answers" [46]. There is, therefore, a need to develop, define and apply economic analyses that not only factor in the positive transaction costs involved in water resources management but also serve as a guiding tool for sustainable water resources management. Subsequent to the 1992 Dublin Conference on Water and Environment, water has widely been accepted as an economic good [47]. Regarding water as an economic good is vital for shifting the behaviour of water users towards efficiency [48]. It is also important for creating a basis for cost recovery related to water access, use and management [48, 49]. However, in defining water as an economic good, there is a need to appreciate the role of water as an environmental, ecological, social, financial and economic resource. A combination of these uses and the multi-sectoral use of water resources have stimulated the consideration that water is complex or "at least very special" [50] compared to other economic goods.

3.7. GIS and remote sensing

To calculate water volume in the basin, the digital elevation model was downloaded from the Japan Aerospace Exploration Agency (Earth Observation Research Centre); http://www.eorc.jaxa.jp/ALOS/en/aw3d30/data/html_v1804/s040e000_s010e030.htm. This dataset has a resolution of approximately 30-meter mesh (1 × 1 arc second). Tiles were downloaded and mosaicked to the study extent (i.e., Botswana, Namibia and Angola). A clipping tool from ArcGIS desktop was used to clip each country to allow easy calculations for the water volume. The surface area tool within the ArcGIS 3D Analyst extension was used to calculate the area and volume of the region, the water surface and the terrain. In order to establish a reference height of the water along the channels of the Okavango Basin, shapefile data for maximum extent of open water for 1984–2015 by Pekel et al. [51] were downloaded, mosaicked, clipped to the basin extent and then converted to a polygon in ArcGIS for desktop. The water extent polygons were then used to extract height values from the digital elevation model. Whereas the original basin DEM had elevations ranging from 695 to 2278 m, the extracted water extent raster showed that the maximum water level reached in the 30-year period used by Pekel et al.

[51] was 1822 m. The mean area of water reached was 7×10^5 km^2, and mean volume reached was 6×10^8 Mm^3. All required datasets were converted into Universal Transverse Mercator projection before calculations were done [52].

Population projection data were taken from OKACOM [10], a shapefile was created and a heat map was created in ArcGIS Pro 2.0 [53] for data visualisation to show density points as a continuous colour gradient.

4. Results and discussion

4.1. Water quantity

Water availability in the ORB is driven by a 30-year multi-decadal variability [5] driven primarily by natural variability [12]. The effect of this variability is more pronounced downstream [5]. This variability translates into years of high floods and years of low floods in the Okavango Delta [5]. Lake level variability in Lake Ngami (in the Okavango Delta) is a good illustration of this multi-decadal variability, where the lake has been more dry than wet between 1880 and 1993 [54, 55]. Therefore, any water developments in the Okavango Delta should account for this multi-decadal variability in water availability.

Arntzen and Setlhogile [18] estimated water abstraction in the ORB at 100 Mm^3, which is less than 1% of the mean annual run-off (MAR). This estimate compares well with FAO [5], who estimated water abstraction in the basin at 90 Mm^3, while water use was estimated at 133 Mm^3. The ORB is a major source of water among the three basin states, and the growing socio-economic needs of an increasing population "are likely to result in greater development of the basin's water resources" [13]. OKACOM [10] revealed that 3428 Ha of land were under irrigation in the basin, while an estimated 490,000 Ha are planned for in Angola, at an estimated water demand of 6400 Mm^3 under future development scenarios. Livestock population in the ORB was estimated at over 1.4 million livestock (i.e., cattle, goats, sheep, etc.), where 67% were in Botswana, of which 65% were cattle [10]. Therefore, the water resource of the ORB is currently underutilised, and water use can be expanded within the limits of the available development space [18]. The two major drivers of water demand in the basin are domestic, industrial and agricultural [16]. Overall, water demand in the ORB is expected to increase gradually to 1857.8 Mm^3 in 2020 to 3871 Mm^3 in 2025, driven mainly by irrigation agriculture [10]. This demand is expected to increase in Angola and Namibia due to water shortages and increased agricultural demand [14].

Water stress is "the ability, or lack thereof, to meet human and ecological demand for water" [56]. According to Gassert [57], the Okavango Basin users have low overall Baseline Water Stress Average Scores of 0.6 [0–5 low–high]. This indicator measures the ratio of total water annual withdrawals to the total available renewable supply, considering upstream uses and depletion of water, and therefore measures the underlying factors that drive water quantity-related risks across basins and countries. Although the overall Okavango Basin's risk scores are much better than the Limpopo and Orange River Basins at 2.7 and 1.9, respectively, all three basins score similarly for interannual variability and flood occurrence. The low basin risk scores are consistent with the *Falkenmark* index (**Table 3**), which suggests that there is

Country	Index ($\times 10^8$)	Category
Angola	83.6	No stress
Botswana	3.4	No stress
Namibia	6.2	No stress
Overall basin	1.8	No stress

Table 3. Summary of the *Falkenmark* index for the ORB states.

no water stress in the basin. Munia et al. [2] also observed that there is no water stress in the ORB either to local or to upstream water uses. This is also in agreement with UNEP-DHI and UNEP [43] global transboundary assessment, which revealed that the ORB has a "very low risk" of transboundary human water stress. Global assessment projections to 2030 indicate no significant changes in human water stress in the basin [43].

An average of 23% of the total discharge in the ORB is needed by the environment to keep it in fair condition [29]. Therefore, this suggests that 2507 Mm^3 of discharge from a total discharge of 10,900 Mm^3 in the basin is needed for ecosystem functioning. Using the mean volume of water available in the basin over a 30-year period (1984–2015) suggests that environmental water needs for the ORB are approximately 1×10^8 Mm^3. Taking into account water withdrawals, it follows that only about 24% of the discharge (or mean volume) in the ORB is utilised for both human use and ecosystem needs. About 76% of the water remains unutilised. Therefore, there is a room for water use expansion in the basin without any concerns of environmental water stress.

4.2. Water quality

Global studies emphasise the need to protect quality of freshwater rivers because they are few, and water demands are exacerbated by the global stressors such as climate change, population growth, industrialisation, economic growth and land-use activities [58]. For some river basins, deteriorating water quality might be as a result of lack of monitoring protocols aimed at protecting the integrity of the wetland in early stages of planning [36]. Therefore, it is critical to continuously monitor water quality in river ecosystems for sustainability of the wetland. The Okavango River water quality is relatively pristine [10, 22], which makes sense given the low human development impact in the basin. The basin's wetland vegetation is partially responsible for the waters' purity, while the substrate of the Kalahari sand also contributes significantly to water quality [10, 17]. Furthermore, wetlands are capable of removing nutrients from surface water resulting in freshwater [59].

Improved surface water quality status of the Okavango Delta is as a result of evapotranspiration and chemical precipitation [60]. Land-use activities such as mining, agriculture, industrialisation and settlements affect the quality of water [61], although slight effects were observed in Maun for pH, total nitrogen and dissolved oxygen [62]. Agricultural activities in the basin result in low nutrient enrichment because of minimal use of fertilisers, herbicides and pesticides. Generally, water quality studies show that waters are aerated and are within the Environmental Protection Agency (EPA) standards [62, 63]. However, there were times when dissolved oxygen (DO) ranged below the minimum of 2.4 mg/L DO as a result of decomposition from organic matter loadings, which resulted in annual fish kills at some parts of the delta [24]. pH levels in the delta are mostly within acceptable guidelines for aquatic life set by EPA and World Health Organisation (WHO), respectively (6.5–9.5 and 6.5–8.5). Few studies looked into major and trace elements and were generally within acceptable limits [63], except for beryllium and aluminium, which exceeded the Botswana Bureau of Standards (BOBS) for drinking water [64]. Although locally the effects of pollution are felt as a result of global stressors, these effects are negligible on the entire delta. Capitalising in good water quality and improved sanitation results in improved human health and economic productivity [65]. Generally, there is localised water pollution around human settlements in the basin [10].

4.3. Ecosystem dynamics

The ORB is internationally important due to its biodiversity and biological productivity [10]. It also contains the Okavango Delta, which is the world largest Ramsar site [4, 13], the world's second largest inland wetland [4] and a key World Heritage Site [66], with a high beta diversity [4]. Due to a low population density, OKACOM [10] revealed that about 90–95% of the basin's natural habitat remained intact. Pröpper et al. [22] also observed that the ORB remains in a "near natural state" due to low development in the basin, where 90% the basin is still covered by natural vegetation. Therefore, ecosystem integrity has remained largely intact because it has remained unaffected by human development due to its remoteness [15]. Therefore, there is high ecosystem connectivity in the ORB, which is a core indicator on transboundary river assessments as described by UNEP-HDI and UNEP [43].

While the ORB is a key source of various natural resources (e.g., firewood, reeds and fruits) for riparian communities, fish is a key source of livelihoods for the basin's communities [17]. According to Ramberg et al. [19], there are approximately 86 fish species in the basin. Fishing is practised on a small-scale commercial in Botswana [15, 67], while it is predominantly artisanal employing crude fishing gear in the rest of the basin [10, 15, 22, 23, 68]. Several studies in the Okavango Delta have shown that the delta's fish stocks are not yet over-exploited [69], where fishing behaviour [70] and fish community dynamics [71, 72] are driven primarily by seasonal flooding. A biodiversity survey in 2012 [23] revealed five previously undescribed fish species in the upper catchment of the ORB, while an earlier survey in 2003 [73] discovered an undescribed fish species in the delta. While earlier studies [19] estimated that there were 86 fish species in the ORB, these recent studies suggest that there are currently about 92 fish species in the system. These observations attest to the pristine status of the Okavango River system as observed by Todd et al. [4].

Despite its pristine status, 330 macro-invertebrate species are either vulnerable or near threatened, 10 fish species are red listed, 3 wetland bird species are considered vulnerable, while another 3 wetland bird species are near threatened in the basin. The common hippopotamus and the African elephant are also considered to be globally vulnerable, but not in the ORB [10] and certainly not in the delta, which have high elephant populations [10, 19]. In fact, the delta was found to be the most productive among several global wetlands due to the seasonal flood pulse [74]. Aquaculture is still at infantile stages in the basin, with small-scale operations in Angola and Namibia [10]. However, it is possible that some entrepreneurs in the upper catchment might decide to farm Nile tilapia, which is fast growing and reaches big sizes, ostensibly to generate higher economic returns. This will have a negative impact on the ORB's ecosystem functioning.

4.4. Governance

Globally, about 145 states share an estimated 286 rivers and lakes and around 200 aquifers [75]. Transboundary freshwater resources are prone to international conflict, especially if not managed properly [76]. These conflicts include economically and verbally hostile actions, which are capable of raising tensions mainly when upstream and downstream interests clash [77]. The institutionalisation of transboundary water management principles can be traced to the Helsinki Rules on the Uses of International Rivers, which are a nonbinding code of conduct for states to follow [78]. They are credited with making the first effort at counter-hegemonic strategies by introducing equity criteria for shared use of international rivers [79]. This then formed the basic foundation for the SADC Protocol [80], which led to the establishment of OKACOM [81]. This agreement replaced the bilateral agreements that existed between the parties [82].

Therefore, OKACOM became a key institution in the basin critical towards management of conflicting interests over common water resources. It sets out rules of the game and mechanisms for the resolutions of potential conflicts. It addresses various issues related to sustainable development, especially on water security and sustainability. In terms of Article 1.4 of the OKACOM Agreement, the commission is an organ responsible for advising the contracting states on the criteria to be adopted for the equitable allocation and sustainable utilisation of water resources in the ORB. Therefore, the commission is obliged to apply principles of "equitable allocation and sustainable utilisation" or the Helsinki Rules [82]. There are, however, several transboundary challenges that OKACOM is faced with. These include lack of harmonisation of water quality standards; insufficient basin-wide cooperation, especially at the local level; and lack of a harmonised land-use planning framework among the basin states, which would facilitate integrated basin planning [10].

Therefore, OKACOM faces both political and structural challenges, which will affect its effectiveness and its ability to ensure that there is sustainable utilisation of the ORB water resources. This will imperil the future of water sustainability in the basin. Pröpper et al. [22] argue that OKACOM is weak and poorly funded, which perhaps attests to lack of political will among the basin states to ensure sustainability of this institution. It is perhaps due to this lack of political will that Pröpper et al. [22] observed that OKACOM recommendations are generally ignored in national decision making among the basin states. Moreover, this lack of effectiveness by OKACOM, underpinned by poor political will, makes the legislative environment unaccommodating to coherent and coordinated policies in the ORB. This observation conforms to Malzbender et al.'s [83] conclusion that there is insufficient integration of environmental policies in the ORB. Generally, this suggests that OKACOM is not fulfilling its mandate as a transboundary river basin management entity. Issues of water sustainability in the future are then potentially imperilled unless the basin states recognise OKACOM's relevance to water resources management in the basin.

4.5. Social systems

The total population in the ORB is approximately 900,000 over 195,000 households and a mean household size of 5. Angola has the smallest household size at 4, while Namibia has the largest household size at 6, and all these depend on the basin for their key livelihood activities [10]. Major livelihood activities in the ORB include subsistence agriculture, harvesting of natural resources, tourism and fishing [10, 15]. The predominant land use in the basin is subsistence agriculture characterised by arable agriculture and pastoral farming, where the largest livestock herds are found in Botswana and Namibia [10]. Flood recession agriculture is common in Botswana and Angola and is relatively more productive than dryland farming [9]. OKACOM [10] further posits that each country has invested in irrigated agriculture, where Namibia has the largest investment in irrigated land, while Angola has planned large future-scale irrigation schemes. Conversely, Botswana has limited irrigation schemes. Harvesting of natural resources includes fisheries, firewood, reeds and grasses, fruits wild foods and medicinal plants. Tourism, whose products are attributable to the wetland, is predominantly non-consumptive and of high value to Namibia and Botswana economies [15]. Aquaculture is yet to reach full potential in all the basin countries [84]. The ORB also contributes to livelihoods through provisioning services [10].

It is evident that most of the basin's communities derive direct use of the basin's resources [15] and that the ORB has a significant contribution to national economies [10]. Anticipated climate change impacts, which include changes in precipitation and temperature, will affect the basin's ecosystem and run-off, which would in turn disturb the availability, allocation and sharing of the ORB water resources [16, 85, 86]. This will inevitably affect the livelihoods of communities that are dependent on these resources. For example, anticipated changes such as loss of habitat, wildlife disruptions, increased wild fires as well as pests and disease vectors would affect the tourism industry [21]. In terms of agriculture, high temperatures will affect both arable and pastoral agriculture, which would force farmers to change the animal breeds and crop varieties [21]. Climate change effects will accentuate poverty in the basin, which is already widespread, and would subsequently increase the vulnerability of communities to socio-economic shocks. Therefore, there is a need to implement adaptation strategies at basin level in order to enhance community resilience. Inadequate basin-wide climate change adaptation and mitigation strategies are a big challenge, while at the same time, there is insufficient long-term policy formulation with respect to climate change adaptation, a common problem in all basin states [10, 83]. Transboundary water resources like the ORB should be protected from potential impacts, even though it is difficult to assess them due to uncertainty inherent in the General Circulation Models (GCMs). This will also fulfil the SDGs, particularly Goal 1 on "no poverty" and Goal 13 on "climate action".

4.6. Water economics and management in the ORB

Factors such as emerging diverse demands for water resources, increasing population, increased urbanisation and rapid evolution of environmental and climatic problems pose future threats to the ORB. It has been projected that the current basin population will increase to about 1.3 million by 2025 [10]. This increase in population will add pressure to the natural water resources. The scarcity of and pressures on water resources in the basin will make the lower basin sharing states vulnerable. This creates interdependencies, which are often perceived as threats [77]. Within the basin, the nation states are faced with attaining a balance between ecosystem function and sustainable resource use by communities dependent on the water resources.

Various modelling techniques for balancing human and ecological water resource needs have been developed [87, 88]. These techniques range from quantifying ecosystem services spatially, to tracing, quantifying and analysing the trade-offs between the ecosystem services. Theoretically, these techniques are needed to inform policy decisions for effective, efficient and secure water resources use and management. Hence, FAO [5] conducted a water audit for the ORB, while Arntzen and Sethogile [18] conducted a water allocation study for the basin, where the goal was to find the balance between human water requirements and ecosystem water needs.

According to FAO [5], a valuation of the Okavango Delta revealed that tourism has the highest direct use value in the delta, the highest contribution to the gross national production and the largest contributor to natural resource rent. Carbon sequestration and wildlife refuge were the largest contributor to indirect use value of the delta [5]. Furthermore, valuation results show that agriculture was less valuable than ecological services in the delta. Annually, the basin's natural resources contribute approximately US$60 million to household income, over US$100 million to the national economy in the form of gross national income and just over US$234 million to the "broad economy in the form of gross national income, including the effect of the national income multiplier" [15]. The largest contributor of this value is tourism activities [5]. However, these economic benefits will reduce basin household income by 50% to approximately US$30 at a low development scenario, to US$10 million household income under a high development scenario [10]. These development scenarios are based on water use in the basin and would mostly involve dams in the upper catchment, which would reduce tourism activities in the lower basin.

One major challenge regarding economic value and benefit in the basin is that Angola is the largest contributor to discharge in the basin but benefits the least, while Botswana contributes the least discharge and derives the largest economic gain [10, 18]. This dichotomy in benefits sharing among basin states is a major transboundary management challenge. Another key challenge in the basin is that while non-consumptive tourism is the most valuable economic activity, agriculture is the biggest source of livelihoods for the majority of households in the basin [5, 18]. This makes water allocation in the basin a challenge. However, the absence of water stress in the basin makes the development space large enough for these activities to occur concurrently. There is a need, however, to intensify data collection in Angola and to enhance management of the water resources of the ORB. This will also provide enough data for development of econometric models as a part of a suite of economic-based models to aid in decision making of water allocation in the basin.

4.7. Climate change

Generally, the ORB is characterised by high inter-annual and multi-annual variability, which makes it resilient to climate variability [13]. GCM's predict increased temperature and decreased rainfall across the ORB with a consequent drop in dry and wet season flows by

49–54% in 2050 and by 68–73% in 2080 [16]. However, Folwell and Farqhuarson [16] conclude that human abstractions will have a minimal impact on both dry and wet season flows. This agrees with OKACOM [10] that water for human abstraction will have a minimal impact on water flows in the system based on future development scenarios.

Generally, climate models consistently project an increase in temperatures over the ORB. The greatest increases in temperatures are associated with high emission scenarios used to force

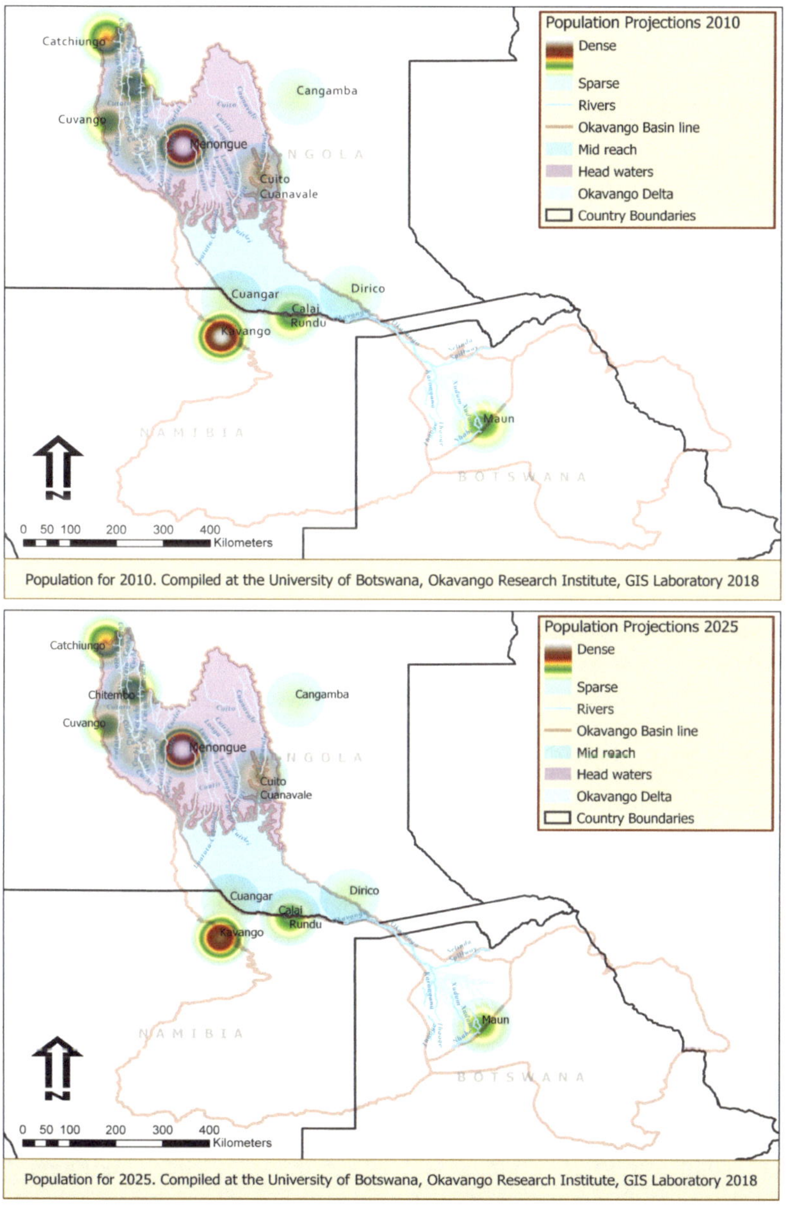

Population for 2010. Compiled at the University of Botswana, Okavango Research Institute, GIS Laboratory 2018

Population for 2025. Compiled at the University of Botswana, Okavango Research Institute, GIS Laboratory 2018

Figure 3. Population growth in key settlements of the ORB.

the models, while the lowest increases are associated with low emission scenarios. For rainfall, the models are inconsistent in projecting the direction of the change (some project wetter, while some project drier future conditions). These predict a negative rainfall change in the future, but the inconsistency in direction of change reduces confidence on the impact of climate change on the ORB. There may also be an increase in heat waves, prolonged dry spells, thunderstorms, localised flooding and damaging winds, which is consistent with other studies (e.g. [86, 89, 90]). The uncertainty in these models makes it difficult to assess the direction of impact, although there is general agreement that the upper basin will receive more rainfall, while the lower basin will receive less rainfall and will experience increased temperatures [18].

4.8. Population growth

Figure 3 shows that there is a relatively higher population density at Menongue and Kavango in 2010 than at Cangamba and Dirico. Population projections for 2025 (**Figure 3**) show that there will not be a major change in population density in the other basin's urban areas, while there will be a negligible increase in population density in Maun. However, this observation is inconsistent with OKACOM [10], which shows that Angola has the highest population growth rate (2.7%) compared to Botswana (0.9%) and Namibia (1%). Furthermore, Angola has the highest fertility rate among the basin states [10]. However, Botswana has the lowest death rate and infant mortality rate among the basin states [10], which may account for an overall negligible increase in Maun population compared to the other urban areas in Angola and Namibia.

5. Synthesis and conclusion

The livelihoods of most of sub-Saharan Africa's rural poor are closely intertwined with water [6], which make water security and water-related issues the most pressing global concerns in the next 10 years [33]. Subsequently, water sustainability becomes a core issue of concern in river basins of developing countries, where future population pressures will conceivably increase water stress in these systems. Lack of preparation to this reality and poor water resources management within this context will invariably lead to fragility, which might trigger simmering tensions among and between the communities [33]. Moreover, comprehensive water resources management is critical towards achieving water security in developing countries. According to Sadoff et al. [33], failure to achieve water security usually occurs through failure by governments to (i) provide its citizens with basic water services, (ii) provide citizens from water-related disasters and (iii) preserve surface, ground and transboundary water resources. Population-driven development pressures will become more of a concern in Angola than downstream states.

Due to low development in the ORB, most of the water needs in the basin is used for ecosystem purposes [91]. Hence, currently, there is no competition for water between human needs and ecosystem requirements. Therefore, this suggests that current consumptive water needs in the system, compared to environmental needs, provide baseline conditions upon which future needs can be assessed against. Clearly, there is a need to conserve the catchment of the Cubango and Cuatir Rivers in the west and the Cuito and Longa Rivers in the east of the

ORB and to ensure that water sustainability is achieved in the basin. Upstream developments, especially dams in these areas, will affect not only water flow dynamics in the entire basin but also sedimentation in the system. However, development pressures in Angola, where 62% of the population is expected to reside by 2025, are realities that need to be addressed within a configured OKACOM. Pröpper et al. [22] argue that multiple uses of ecosystem services in the ORB (e.g., agriculture, fish, water supply, etc.) will possibly result in over-utilisation and commodification with unknown consequences on the ecosystem. Invariably, this may affect water sustainability in the basin.

Yang's et al. [92] six major water management strategies for the Texas State Water Plan can also be applied to the ORB to enhance water resources management in the basin and hence increase water sustainability. These include "water conservation, surface water development, groundwater development, re-use, desalination and conjunctive use" [92]. Similar to the Texas situation, it is envisaged that implementing some of these management strategies in the river basin will in the long term alleviate pressure on surface water resources, especially for uses other than ecosystem needs. Ultimately, this will release more water for ecosystem functioning. Some of these strategies, especially re-use and water conservation, can be codified into law to ensure that they become part of water governance in the ORB. Moreover, OKACOM should be the key driver of this envisaged policy formulation among the basin states. There is no harmonised policy framework in the ORB that deals with water legislation among the basin states [83].This is a major weakness of the OKACOM policy framework because this suggests that there is potentially lack of coordination in water resources management among the basin states, which might result in unsustainable water use.

The water allocation problem for ecosystem needs on one hand and for human livelihoods on the other hand will increasingly become more of a political issue than an economic activity in future [3]. This observation is premised on the fact that increasing population size in the ORB will undoubtedly place more pressure on water resources in the basin, and basin states will increasingly be faced with the political pressures of providing services to their populations, at the expense of ecosystem needs. Currently, however, Pröpper et al. [22] observe that economic pressures among the ORB states are currently the key drivers of water resources management policies. Nonetheless, the basic question then remains, how do we achieve water sustainability in the ORB as we move into the future? Currently, OKACOM is ill equipped to deal with these political questions that underpin water use and allocation in the basin, in the face of the inevitable increasing population pressures. What implementable and practical measures have been implemented in the basin states to deal with these future threats?

Water is the basic foundation of the SDGs and is hence the key determinant of their successes [42]. There is a need to ensure that the water resources of the ORB are used sustainably, through finding a balance between ecosystem needs and human livelihoods. This will ensure that the basin states achieve the SDGs. Future major potential threats on the basin's water resources are issues related to water quality and habitat fragmentation caused by dam construction in the upper catchment. While these future development scenarios will cause an overall reduction in ORB household income, this will be related primarily to loss of tourism activities in the delta. Therefore, OKACOM needs to create a decision support system (DSS) to facilitate benefit sharing in the basin. Hopefully, this will assist the basin states to find the balance between ecosystem dynamics and human livelihoods, which will reduce or eliminate competition for water requirements. This will contribute to water sustainability in the ORB.

Author details

Ketlhatlogile Mosepele[1*], Wame L. Hambira[1], Goemeone E.J. Mogomotsi[2], Patricia K. Mogomotsi[1], Oliver Moses[1], Masego Dhliwayo[1], Anastacia Makati[1] and Bernice Setomba[1]

*Address all correspondence to: kmosepele@ori.ub.bw

1 Okavango Research Institute, University of Botswana, Maun, Botswana

2 University of Botswana, Gaborone, Botswana

References

[1] Rojas GA, Iza A. Governance of Shared Waters. Legal and Institutional Issues. Gland: IUCN; 2011. p. x + 230

[2] Munia H, Guillaume JHA, Mirumachi N, Porkka M, Wada Y, Kummu M. Water stress in global transboundary river basins: Significance of upstream water use on downstream stress. Environmental Research Letters. 2016;**11**(1):014002. DOI: 10.1088/1748-9326/11/1/014002

[3] Loucks DP, Haifeng JIA. Managing water for life. Frontiers of Environmental Science & Engineering. 2012;**6**(2):255-264. DOI: 10.1007/s11783-011-0359-6

[4] Todd MC, Andersson L, Ambrosino C, Hughes D, Kniveton DR, Mileham L, et al. Climate change impacts on hydrology in Africa: Case studies of River Basin water resources. In: Williams CJR, Kniveton DR, editors. African Climate and Climate Change, Advances in Global Change Research 43. Dordrecht, Netherlands: Springer Science; 2011. pp. 123-153. DOI: 10.1007/978-90-481-3842-5_6

[5] FAO. AQUASTAT Main Database. Food and Agriculture Organization of the United Nations (FAO) [Internet]. 2014. [Accessed: Jun 14, 2018]

[6] World Bank. Climate Resilience in Africa: The Role of Cooperation Around Transboundary Waters. Washington, DC: World Bank; 2017. 110 p

[7] Junk JJ. Long-term environmental trends and the future of tropical wetlands. Environmental Conservation. 2002;**29**(4):414-435. DOI: 10.1017/S0376892902000310

[8] Kgathi D, Kniveton D, Ringrose S, Turton R, Vanderpost C, Lundqvist J, et al. The Okavango: A river supporting its people, environment and economic development. Journal of Hydrology. 2006;**331**:3-17

[9] Weinzier T, Schilling J. On demand, development and dependence: A review of current and future implications of socioeconomic changes for integrated water resource management in the Okavango Catchment of Southern Africa. Land. 2013;**2**(1):60-80

[10] OKACOM. The Permanent Okavango River Basin Water Commission: Cubango-Okavango River Basin Transboundary Diagnostic Analysis. Maun: OKACOM; 2011. 219 p

[11] van Rijswick M, Edelenbos J, Hellegers P, Kok M, Kuks S. Ten building blocks for sustainable water governance: An integrated method to assess the governance of water. Water International. 2014;**39**(5):725-742. DOI: 10.1080/02508060.2014.951828

[12] Wolski P, Stone D, Tadross M, Wehner M, Hewitson B. Attribution of floods in the Okavango basin, Southern Africa. Journal of Hydrology. 2014;**511**:350-358

[13] Hughes DA, Kingston DG, Todd MC. Uncertainty in water resources availability in the Okavango River Basin as a result of climate change. Hydrology and Earth System Sciences. 2011;**15**:931-941. DOI: 10.5194/hess-15-931-2011

[14] Andersson L, Gumbricht T, Hughes D, Kniveton K, Ringrose S, Savenije H, et al. Water flow dynamics in the Okavango River Basin and Delta—A prerequisite for the ecosystems of the Delta. Journal of Hydrology. 2003. DOI: 10.1016/j.jhydrol.2006.04.039

[15] Barnes J, Saraiva R, Mmopelwa G, Mbaiwa J, Magole L, Wamunyima D. Okavango River Basin Transboundary Diagnostic Analysis: Socio-Economic Assessment Final Report. Maun: OKACOM; 2009

[16] Folwell S, Farqhuarson F. The impacts of climate change on water resources in the Okavango basin. In: Proceedings of the Fifth FRIEND World Conference on Climate Variability and Change—Hydrological Impacts. Havana, Cuba: IAHS Publ.; Nov 2006. p. 308

[17] Mendelsohn JM, vanderPost C, Ramberg L, Murray-Hudson M, Wolski P, Mosepele K. Okavango Delta: Floods of Life. Windhoek: RAISON; 2010. 144 p

[18] Arntzen J, Setlhogile T. Towards an Water Allocation Strategy for the Cubango Okavango River Basin. Gaborone: CAR; 2015. 25 p

[19] Ramberg L, Hancock P, Lindholm M, Meyer T, Ringrose S, Sliva J, et al. Species diversity of the Okavango Delta, Botswana. Aquatic Science. 2006;**68**:310-337

[20] Soderstrom E, Purkey D, Matiza T, Ashton P, Madzwamuse M, Walking R, et al. Transboundary Collaborative Learning: Case Study in the Okavango River Basin; Comprehensive Assessment Project Report (Draft). International Water Management Institute (IWMI) and USAID [Internet]. 2006. Available from: http://www.iwmi.cgiar.org/assessment/FILES/word/ProjectDocuments/Okavango/Okavango_Draft%20Report.pdf [Accessed: Nov 28, 2008]

[21] SAREP. Cubango-Okavango River Basin Climate Change Vulnerability Assessment, Technical Series. Vol. 1a. Washington, DC: Chemonics International; 2014. 43 p

[22] Pröpper M, Gröngröft A, Finckh M, Stirn S, De Cauwer V, Lages F, et al. The Future Okavango—Findings, Scenarios and Recommendations for Action. Research Project Final Synthesis Report 2010-2015. 190 p

[23] SAREP. Biodiversity Survey of the upper Angolan Catchment of the Cubango-Okavango River Basin. Washington, DC: Chemonics International; 2012. 151 p

[24] Alonso LE, Nordin LA. A Rapid Biological Assessment of the Aquatic Ecosystems of the Okavango Delta, Botswana: High Water Survey. Washington, DC: Conservation International; 2004

[25] UNEP-DHI and UNEP. Transboundary River Basins: Status and Trends, Summary for Policy Makers. Nairobi: United Nations Environment Programme (UNEP); 2016. 12 p

[26] Cook SE, Fisher MJ, Andersson MS, Rubiano J, Giordano M. Water, food and livelihoods in river basins. Water International. 2009;**34**(1):13-29. DOI: 10.1080/02508060802673860

[27] Brown A, Marty D. Matlock. Review of Water Scarcity Indices and Methodologies. White Paper #106. The University of Arkansas: The Sustainability Consortium; 2011. 21 p

[28] Forslund A et al. Securing Water for Ecosystems and Human Well-being: The Importance of Environmental Flows. Stockholm: Swedish Water House Report 24. 2009. 52 p

[29] Smakhtin V, Revenga C, Döll P. A pilot global assessment of environmental water requirements and scarcity. Water International. 2004;**29**(3):307-317

[30] Masocha M, Murwira A, Magadza CHD, Hirji R, Dube T. Remote sensing of surface water quality in relation to catchment condition in Zimbabwe. Physics and Chemistry of the Earth. 2017;**100**:13-18

[31] Wongsupap C, Weesakul S, Clemente R, Das Gupta A. River basin water quality assessment and management: case study of Tha Chin River Basin, Thailand. Water International. 2009;**34**(3):345-361. DOI: 10.1080/02508060903115134

[32] Hughes WB, Abrahamsen TA, Maluk TL, Reuber EJ, Wilhelm LJ. Water Quality in the Santee River Basin and Coastal Drainages, North and South Carolina, 1995-98. Circular 1206. Washington, DC: U.S. Geological Survey; 2000. 40 p

[33] Sadoff CW, Borgomeo E, de Waal D. Turbulent Waters: Pursuing Water Security in Fragile Contexts. Washington, DC: World Bank; 2017. 61 p

[34] Cools J, Johnston R, Hattermann FF, Douven W, Zsuffa I. Tools for wetland management: Lessons learnt from a comparative assessment. Environmental Science and Policy. 2013;**34**:138-145

[35] Sun R, Yao P, Wang W, Yue B, Liu G. Assessment of wetland ecosystem health in the Yangtze and Amazon River Basins. International Journal of Geo-Information. 2017;**6**:81. DOI: 10.3390/ijgi6030081

[36] Ramsar. River basin management: Integrating wetland conservation and wise use into river basin management. Ramsar Handbooks for the Wise Use of Wetlands. 4th ed. Vol. 9. Gland: Ramsar Convention Secretariat; 2010

[37] Ostrovskaya E, Douven W, Schwartz K, Pataki B, Mukuyu P, Kaggwa RC. Capacity for sustainable management of wetlands: Lessons from the WETwin project. Environmental Science and Policy. 2013;**34**:128-137

[38] Namaalwa S, Van damb AA, Funk A, Ajie GS, Kaggwa RC. A characterization of the drivers, pressures, ecosystem functions and services of Namatala wetland, Uganda. Environmental Science and Policy. 2013;**34**:44-57

[39] Arias-Hidalgo M, Villa-Cox C, Griensven AV, Solo'rzano G, Villa-Cox R, Mynett AE, et al. A decision framework for wetland management in a river basin context: The "Abras de Mantequilla" case study in the Guayas River Basin, Ecuador. Environmental Science and Policy. 2013;**34**:103-114

[40] Tortajada C. Water governance: Some critical issues. International Journal of Water Resources Development. 2010;**26**(2):297-307. DOI: 10.1080/07900621003683298

[41] Orme M, Cuthbert Z, Sindico F, Gibson J, Bostic R. Good transboundary water governance in the 2015 sustainable development goals: A legal perspective. Water International. 2015;**40**(7):969-983. DOI: 10.1080/02508060.2015.1099083

[42] World Bank. High and Dry: Climate Change, Water, and the Economy. Washington, DC: World Bank; 2016. 69 p

[43] UNEP-DHI and UNEP. Transboundary River Basins: Status and Trends. Nairobi: United Nations Environment Programme (UNEP). 2016. 378 p

[44] Challen R. Institutions, Transaction Costs, and Environmental Policy: Institutional Reform for Water Resources. Cheltenham: Edward Elgar Publishing Limited; 2000

[45] Saleth RM, Dinar A. A Cross-Country Analysis of Institutions and Performance. Cheltenham: Edward Elgar Publishing Limited; 2004

[46] Coase R. The Firm, The Market and the Law. Chicago: University of Chicago Press; 1988

[47] Savenije H, van der Zaag P. Water as an economic good and demand management: Paradigms with pitfalls. Water International. 2002;**27**(1):8-104

[48] Yuling S, Lein H. Treating water as an economic good: Policies and practices in irrigation agriculture in Xinjiang, China. The Geographical Journal. 2010;**176**(2):124-137

[49] Madigele P. The complexity of the economic conception of water: Revisiting water resources management in developing countries. In: Mmopelwa G, editor. Water Resource Management; Science and Technology Innovation for Sustainable Development. Calgary: Acta Press; 2016. pp. 87-92

[50] Savenije H. Why water is not an ordinary economic good, or why the girl is special. Physics and Chemistry of the Earth. 2002;**27**:741-744

[51] Pekel JF, Cottam A, Gorelick N, Belward AS. High-resolution mapping of global surface water and its long-term changes. Nature. 2016;**540**:418

[52] Xiaoyan Y, Li L, Longgao C, Longqian C, Shen Z. Improving ASTER GDEM accuracy using land use-based linear regression methods: A case study of Lianyungang, East China. ISPRS International Journal of Geo-information. 2018;**7**(4):145

[53] Keranen K, Kolvoord R. Making Spatial Decision using ArcGIS Pro: A Workbook. Redlands: ESRI Press; 2017

[54] Shaw PA. Fluctuations in the level of Lake Ngami: The historical evidence. Botswana Notes and Records. 1983;**15**:79-84

[55] Shaw P. The desiccation of Lake Ngami: An historical perspective. The Geographical Journal. 1985;**151**:318-326

[56] CEO Water Mandate. Driving Harmonization of Water Stress, Scarcity, and Risk Terminology. Discussion Paper, UN Global Impact. 2014

[57] Gassert F, Reig P, Luo T, Maddocks A. Aqueduct Country and River Basin Rankings: A Weighted Aggregation of Spatially Distinct Hydrological Indicators. Washington, DC: World Resources Institute; 2013

[58] Padowski JC, Gorelick SM, Thompson BH, Rozelle S, Fendorf S. Assessment of human-natural system characteristics influencing global freshwater supply vulnerability. Environmental Research Letters. 2015;**10**(10):104014

[59] Verhoeven JT, Arheimer B, Yin C, Hefting MM. Regional and global concerns over wetlands and water quality. Trends in Ecology and Evolution. 2006;**21**(2):96-103

[60] McCarthy TS. Groundwater in the wetlands of the Okavango Delta, Botswana, and its contribution to the structure and function of the ecosystem. Journal of Hydrology. 2006;**320**:264-282

[61] Meck M, Love D, Mapani B. Zimbabwean mine dumps and their impacts on river water quality a reconnaissance study. Physics and Chemistry of the Earth. 2006;**31**:797-803

[62] Masamba WR, Mazvimavi D. Impact on water quality of land uses along Thamalakane-Boteti River: An outlet of the Okavango Delta. Physics and Chemistry of the Earth, Parts A/B/C. 2008;**33**(8-13):687-694

[63] West DT, van As JG, van As LL. Surface water quality in the Okavango Delta panhandle, Botswana. African Journal of Aquatic Science. 2015;**40**(4):359-372. DOI: 10.2989/16085914.2015.1104288

[64] Mogobe O, Mosimanyana E, Masamba WRL, Mosepele K. Monitoring water quality of the Upper Okavango Delta. In: Revermann R, Krewenka KM, Schmiedel U, Olwoch JM, Helmschrot J, Jürgens N, editors. Climate Change and Adaptive Land Management in Southern Africa—Assessments, Changes, Challenges, and Solutions, Biodiversity and Ecology. Vol. 6. Maun: Klaus Hess Publishers; 2018. pp. 106-111. DOI: 10.7809/b-e.00311

[65] Millennium Ecosystem Assessment. Millennium ecosystem assessment. In: Ecosystems and Human Well-Being: Biodiversity Synthesis. Washington, DC: World Resources Institute; 2005

[66] Matswiri GM. Two in one: Explaining the management of the Okavango Delta World Heritage Site, Botswana [thesis]. Cape Town: University of Cape Town; 2017

[67] Mosepele K, Ngwenya B. Socio-Economic Survey of Commercial Fishing in the Okavango Delta, Botswana. Gaborone: Bay Publishing; 2010

[68] Bills R, Mazungula N, Almeida F. A survey of the fishes of upper Okavango river system (Cuito and Kwando rivers) in Angola. SAIAB Investigational Report No. 2013. 25 p

[69] Mosepele K, Kolding J. Fish stock assessment in the Okavango delta, Botswana—Preliminary results from a length based analysis. In: Bernard T, Mosepele K, Ramberg L, editors. Environmental Monitoring of Tropical and Subtropical Wetlands. Maun: University of Botswana; 2003. pp. 363-390

[70] Mosepele K, Mmopelwa G, Mosepele B, Kgathi DL. Indigenous knowledge and fish utilization in the Okavango Delta, Botswana: Implications for food security. In: Ahmed A, editor. Managing Knowledge, Technology and Development in the Era of Information Revolution. Australia: Griffith University; 2007. pp. 292-302

[71] Mosepele K, Moyle PB, Merron GS, Purkey D, Mosepele B. Fish, floods and ecosystem engineers; interactions and conservation in the Okavango Delta, Botswana. Bioscience. 2009;**59**(1):53-64

[72] Mosepele K, Kolding J, Bokhutlo T. Fish community dynamics in an inland floodplain system of the Okavango Delta, Botswana. Ecohydrology and Hydrobiology. 2017;**17**:89-102. DOI: 10.1016/j.ecohyd.2017.01.005

[73] Tweddle D, Bills R, van der Waal B, Skelton P, Kolding J, Nengu S. Fish diversity and fisheries in the Okavango Delta, Botswana. In: Alonso LE, Nordin L, editors. A Rapid Biological Assessment of the Aquatic Ecosystems of the Okavango Delta, Botswana: High Water Survey. RAP Bulletin of Biological Assessment 27. Washington, DC: Conservation International; 2003. pp. 97-110

[74] Junk WJ, Brown M, Campbell IC, Finlayson M, Gopal B, Ramberg L, et al. The comparative biodiversity of seven globally important wetlands: A synthesis. Aquatic Sciences. 2006;**68**:400-414. DOI: 10.1007/s00027-006-0856-z

[75] Wolf AT, Natharius JA, Danielson J, Ward BS, Pender JK. International River Basins of the world. International Journal of Water Resources Development. 1999;**15**(4):387-427

[76] Haftendorn H. Water and international conflict. Third Word Quarterly. 2000;**21**(1):51-68

[77] Petersen-Perlman JD, Veilleux JC, Wolf AT. International water conflict and cooperation: Challenges and opportunities. Water International. 2017;**42**(2):105-120

[78] Gupta J. The watercourses convention, hydro-hegemony and transboundary water issues. The International Spectator. 2016;**51**(3):118-131

[79] Benvenisti E. Collective action in the utilization of shared freshwater: The challenges of international water resources law. American Journal of International Law. 1996; **90**(3):384-415

[80] Rieu-Clarke A, Kinna R. Can two global UN water conventions effectively co-exist? Making the case for a 'package approach' to support institutional coordination. Review of the European Community and International Environmental Law. 2014;**23**(1):15-31

[81] Marcantonio RA. The Okavango River Basin agreement: An analysis. African Geographical Review. 2016;**35**(3):227-246

[82] Mogomotsi GE, Mogomotsi PK, Mosepele K. Legal aspects of transboundary management framework of the Okavango River Basin. Leiden Journal of International Law (forthcoming)

[83] Malzbender D, Mokorosi P, Jooste D, Tweedy H. Environmental Protection and Sustainable Management of the Okavango River Basin (EPSMO) Governance Review. Okavango River Basin Transboundary Diagnostic Report. Maun: OKACOM; 2010. 66 p

[84] Domptail S, Mundy O. VISIONS 2030 for the Okavango River Basin: Inspirations from 27 Stakeholders from All Levels Learning from One another to Better Anticipate the Future. Giessen: Universitet Giessen; 2014. 40 p

[85] Khedun P, Sanchez R, Giardino JR. Climate change transboundary water resources. Water Resources Impact. 2009;**11**(2):11-13

[86] IPCC Summary for policymakers. In: Field CB, Barros VR, Dokken DJ, Mach KJ, Mastrandrea MD, Bilir TE, et al., editors. Climate Change 2014: Impacts, Adaptation, and Vulnerability. Part A: Global and Sectoral Aspects. Contribution of Working Group II to the Fifth Assessment Report of the Intergovernmental Panel on Climate Change. Cambridge: Cambridge University Press; 2014. pp. 1-32

[87] Cleveland C, Costanza R, Eggertsson T, Fortmann L, Low B, McKean M, et al. A Framework for Modeling the Linkages between Ecosystems and Human Systems. Stockholm: Beijer Discussion Paper Series No. 76. 1996

[88] Postel S, Richter B. Rivers for Life—Managing Water for People and Nature. Washington, DC: Island Press; 2003

[89] Wolski P. Trends in Water Resources Base in Okavango River Basin. Rome: FAO; 2012

[90] Moses O. Climate change in Lake Ngami, Botswana. In: Kgathi DL, Mosimanyana EN, Mbaiwa JE, Masamba WRL, editors. Ecosystem Services and Human Well-being at Lake Ngami, Botswana. New York: Nova; 2018. pp. 61-69

[91] Green OO, Cosens BA, Garmestani AS. Resilience in transboundary water governance: The Okavango River Basin. Ecology and Society. 2013;**18**(2):23. DOI: 10.5751/ES-05453-1802232013[95

[92] Yang Y, Pi Z, Lu Q, Austin B. Assessing the effects of water management strategies on streamflow using water availability models. Water International. 2009;**34**(3):335-344. DOI: 10.1080/02508060903114673

Chapter 5

Sustainability of Irrigation in Uzbekistan: Implications for Women Farmers

Elena Kim

Additional information is available at the end of the chapter

http://dx.doi.org/10.5772/intechopen.79732

Abstract

This chapter focuses on a discussion of how global efforts to align local irrigation management with the good governance principles affect the lives of the rural poor, specifically women. Drawing in empirical data collected in post-soviet Uzbekistan, I illuminate unexpected effects of an apparently well-intended irrigation project on those categories of farmers whose connections to state apparatus of agricultural commerce of cotton were weak. Using fieldwork data from a village largely affected by desiccation of Aral Sea, I describe the everyday struggles by these people, who are mostly women, engage to make their living and provide subsistence to their families in situation of economic trauma, environmental disaster, and massive outmigration of male population. This analysis puts forward the local voices of real people whose lives are being restructured by sustainability oriented actions. Such perspective is often missed in scholarly and professional literature. These findings are hoped to assist policy developers in formulating irrigation programs in ways that would embrace sustainability both in terms of environmental and social justice.

Keywords: irrigation, gender, women farmers, sustainability, Aral Sea

1. Introduction

This chapter focuses on a discussion of how global efforts to align national irrigation management with the good governance principles affect the lives of the rural poor, specifically women. Drawing in empirical data collected in post-soviet Uzbekistan, I illuminate unexpected effects of an apparently well-intended project aimed at establishing sustainable water practices on those categories of farmers whose connections to state apparatus of agricultural commerce of cotton were weak. Using fieldwork data from a village largely affected by desiccation of

Aral Sea, I describe the everyday struggles these people, who are mostly women, engage to make their living and provide subsistence to their families in situation of economic trauma, environmental disaster, and massive out-migration of male population. This analysis puts forward the local voices of real people whose lives are being restructured by sustainability oriented actions. Such perspective is often missed in scholarly and professional literature. These findings are hoped to assist policy developers in formulating irrigation programs in ways that would embrace sustainability both in terms of environmental and social justice.

2. Water politics in Uzbekistan

Contemporary Uzbekistan, or formally, Republic of Uzbekistan, is one of the independent post-Soviet Central Asian States. It was created in early 1930s, when the decision of Moscow defined its geographic boundaries, as well as those of other several republics, on the territory which was then called the Soviet Turkestan [1]. The political objective of that time was to boost Soviet Union's economic performance within which the Soviet Uzbek Republic acquired a specialization in the political economic system of the Soviet Union—it supplied cotton for Russian textile mills [1]. Cotton production mediated important links between Moscow, local elites, and regional elites which altogether ensured that the cotton production plan was fulfilled [2] in exchange for additional financial transfers from Moscow [2].

Uzbekistan's cotton agriculture was based on the irrigation where the main canal feeding the farms had been constructed in Fergana Valley three centuries earlier and supplemented with 1960s irrigation infrastructure built to draw vast amounts of irrigation water from the region's two major rivers, Amu Darya and Syr Darya [3]. The Soviet planners insistent on maximum cotton output and the country engaged in an intensive monoculture to foster cotton production through expansion of irrigated areas. Between 1960 and 1990, the irrigated areas in the country increased by 2 million hectares (about 60% of all irrigated land in Central Asia) [2]. The Soviet administration enforced a specific organization of agricultural work through massive collectivization to consolidate individual landholding and labor into collective farms called "kolkhoz" (collectively owned) and "sovkhoz" (state owned) forms of agricultural production cooperatives [4]. In a kolkhoz, the workers received a share of the farm's product and profit according to the number of days worked. In a sovkhoz, the workers received a fixed salary. The Soviet state administration developed and imposed work programs for these collective farms and nominated their preferred managers (brigadirs). By 1990, Uzbekistan had about 940 kolkhoz and more than a 1000 of sovkhoz [5]. In mid-1990, the Soviet Uzbek Republic adopted a Law on Land, which allowed individual to hold land for private plots and individual farms on long-term lease [5]. The law permitted tenure to be inherited but did not permit agricultural land to be privately owned.

In 1991, the USSR collapsed and Uzbekistan gained sovereignty. The Government of independent Uzbekistan initiated a policy of transition from the Soviet centrally planned economy to a market economy [6]. Farm restructuring was one of the major components of the transition agenda with land management and water reforms that ensued. Policy shifts took place in the context where the trading links with other republics which were previously orchestrated by Moscow were now disrupted and resulted in shortfall of grain. Dissolution of collective farms led to massive unemployment and livelihood insecurity among the population, especially in the rural areas [2]. The Uzbek government responded to the shortages by expanding the acreage of land devoted to wheat production and increased the size of private plots that population became entitled to [2]. The agrarian reform oscillated between increasing access to private land, structural reform agenda imposed by international donors, and measures to tighten and restrict private access to land in an effort to control the production of cotton [2].

Agrarian reforms transformed collective farms to collective enterprises, then, again, restructured them as joint-stock companies and, lastly, established private enterprises such as independent farms [7]. The private farms were made distinct from peasant farms in that they had a legal status, had a leasehold of up to 50 years, had a minimum of 10 hectares for cotton and wheat, and their land use was restricted to specific agricultural activities as specified in the lease contracts [2]. The peasant farms had optional legal status, had a life-long inheritable tenure, could only use family members and relatives as labor, had a maximum size up to 1 ha, and might use their land for any agricultural activities. The private farms were the subject to a mandatory system of production quotas and state orders on production of cotton and wheat [6]. Prices were fixed by the government-controlled agencies and well below the market prices. The state used a system of contracting private farmers, whereby they became bound to continue to plant a certain acreage of cotton [2]. Should they fail to supply the expected amount, the producers were subject to punitive measures such as revoking their leases. In return, producers were supplied with rationing of inputs such as land, water, equipment, etc. As Kandiyoti [2] argued, this was an attempt of the government to pass on the production risks to the independent farmers, while maintaining the state control over the procurement of strategic crops such as cotton and wheat. The small holders endured no state demands aside from land tax.

As in Kandiyoti [2], Uzbekistan's agrarian reform systematically disadvantaged women. For example, when the members of collective farms were redefined as shareholders, women received much smaller shares than men because those were distributed on the basis of the length of service and final salaries. Women, most of them were unskilled workers with shorter working years and frequent maternity leaves fared considerable less than men. The notion that farms were to be managed by men was becoming a fact.

The importance of land for sustaining rural livelihoods also underwent changes. In contrast to the Soviet period, where individual holding did not play a significant role, in sovereign Uzbekistan, subsistence and informal income from individual crop cultivation became central to families surviving strategies [8]. When waged employment became permanently deficit, state benefits became irregular and curtailed, reliance on households and subsidiary plots for self-subsistence increased substantially, and rural households turned to self-provisioning and sale or barter of produce [2].

In contemporary Uzbekistan, agriculture accounts to 30% of GPD, 60% of foreign exchange receipts, and about 40% of employment [9]. Private farms carry on producing cotton for the state international commerce and make Uzbekistan now the world's fourth largest producer of cotton [10]. This happens despite the fact that the Uzbekistan currently suffers a serious water shortage [10]. The country continues to use the same irrigation sources and infrastructure, mainly from Syr Darya and Amu Darya rivers, which feed the landlocked Aral Sea which used to be one of the world's largest saline lakes. But when the Soviet government decided to divert these rivers to irrigate the desert areas surrounding the Sea to supply irrigation to agriculture, the inflow was reduced from more than 50 cubic kilometers of water in 1960, to 42.5 in 1973, 8.3 by 1980, and 0 by 1982 [5].

The Aral Sea started to desiccate, and in later 1990s, its water level was only one-fifth of what it used to be four decades earlier [3]. By 2007, the Aral Sea had shrunk to 10% of its original size [3, 11]. Today, it is recognized as one of the largest environmental disasters, an environmental, social, and economic tragedy which poses environmental, social, economic, humanitarian, public health, and other risks [11]. Today, people living around the Aral Sea (about 60 million people) are some of the poorest in Central Asia and suffer declining fresh water supply, pollution, violent sand storm, and public health risks [11].

Beginning in early 1990, global communities drew their attention to the Aral Sea problem and its insinuating links with regional and global security issues. At the 48th and 50th session of the UN General Assembly on September 28, 1993 and on October 1995, Central Asian Delegation appealed to the global community to help save the Aral Sea. In 2010, the UN Secretary General, Ban Ki-Moon called the Aral Sea crisis "clearly one of the worst environmental disasters in the world" [12]. He urged the central Asian leaders to cooperate and seek for solutions and promised that "all specialized agencies of the United Nations will provide necessary assistance and expertise" [12]. A number of UN programs were initiated to improve economic, food, health, and environmental security among the poor rural communities of the Aral Sea-affected areas. Under the auspices of the United Nations, in September 2015, the government of Uzbekistan initiated the establishment of a Trust Fund for the Aral Sea. In 2017, the current UN Secretary General, Antonio Guterres, visited the Aral Sea basin and made a public statement about it expressing his concerns and calling for remedial action [13].

Since 1991, many water-related programs and projects were implemented by the international donor communities and made significant contributions to the agricultural sector in Uzbekistan through infrastructure rehabilitation, installation of water monitoring systems, etc. Funding agencies included Asian Development Bank, Global Environmental Facility, Swiss Agency for Development and Cooperation, the World Bank, Food and Agriculture Organization, United Nations Development Programme, USAID, the German Agency for International Cooperation, European Union, Japan International Cooperation Agency, etc. One of the biggest projects was the Asian Development Bank's Amu Bukhara Irrigation System Rehabilitation project with funding of US$ 320 million. All of these projects introduced important changes in the management of irrigation and fresh water, built local capacity, and produced large research-based knowledge; however, a short survey of these project demonstrates while the project self-describe themselves as successful, and they also demonstrate lack of attention to water-scarce regions of Uzbekistan and intra-project coordination and cooperation.

3. Sustainability project

Instigated by the 1992 United Nations Conference on Environment and Development (UNCED) in Rio de Janeiro and its resultant international conventions on climate change (UNFCCC), biodiversity (UNCBD) and desertification (UNCCD), some of the European countries started expressing interests in joining cooperation to address the Aral Sea crisis. In this chapter, the focus is on the German Government's response to the challenges posed by these three conventions and their importance for its national research and development strategy. In mid-1990s, The German Development Cooperation, a large government-sponsored worldwide organization for international cooperation, announced among its priorities combating desertification of Central Asian region through efficient water and land use in the Aral Sea region (in its report to the UN Secretariat for the UNCCD) [14]. Since 1992, German the Federal Ministry for Education and Research (BMBF), the German largest governmental funding agency that takes an overarching responsibility for science and research-related policies [15], BMBF cooperated with UNESCO providing funding for projects with the aim to assess and respond to damage to the Aral Sea's ecosystems. The first such project took place in 1993–1999 with a budget of US$ 1.2 million and supported the network of 140 scientists from Central Asia and Russia who worked on 20 various subprojects.

In 2001, BMBF proposed the second project called "Economic and Ecological Restructuring of land and Water Use in the Region Khorezm, Uzbekistan", worth US$10 million which lasted from 2001 to 2010. The Center for Development Research at Bonn University in Germany developed and implemented it in close cooperation with UNESCO and Urgench State University in Uzbekistan. The project (henceforth, BMBF-UNESCO project) aimed at addressing environmental, social, and economic problems in Khorezm, one of the three provinces in the Aral Sea zone "in the context of the Aral Sea crisis to provide sound, science-based policy recommendations for sustainably improving the natural resource use" ([14] p. 6). Part of the project focused on rational water and land use via a shift toward sustainable land and water resource management. It was expected that sustainable resource use would lead to more efficient agriculture and improve rural livelihoods. This project is the focus of this chapter where I show complexity of operationalizing the concept of sustainability in specific projects and programs.

The BMBF-UNESCO project was subdivided into phases. The first two phases were implemented in 2001–2006 with an overall goal to develop region-specific innovative technologies in land and water use via scientific modeling. During phases I and II, project scientists compiled databases and completed baseline investigations of groundwater and soil salinity, estimated water budgets for regional irrigations, assessed soil conservation agriculture, etc. On the basis of this knowledge, a number of the so-called "plausible solutions" were selected to be applied in real-life settings during phase 3. Phase 3 explicitly provided space for social issues within technical options for sustainable land and water use management. Phase 3 planners committed to notions of sustainability on the basis of participation, bottom-up approaches, and improved rural livelihoods. In one of the project components, it was envisioned that sustainable water use could be achieved by introducing community-based water management through an improved operation of the existing WUA. This was one of the promising solutions, an innovation to be followed. The idea was that target groups participated in testing the innovations together and used them independently once the proposed solutions proved suitable and sustainable. The work on WUA started in 2008 under the name "Social mobilization and institutional development (SMID)", whereby the project was to improve the local WUA which had been assessed as "weak." This work was expected to improve "the livelihoods of rural inhabitants and enhance productivity of the irrigated agriculture through better water management" ([20] p. 5). The SMID approach relied on two major directions which were seen as appropriate for attaining the envisioned goal. One component of the work called Social Mobilization aimed at making the WUA known and understood by the villagers in order to generate "ownership, social, monetary and labor support from the water users to the WUA" ([21] p. 1) to its WUA and an overall wider "inclusion of the large share of water users and their concerns into the decision making processes of the WUA" ([20] p. 1). The second direction was called Institutional Development which stressed the importance of WUA's organizational growth as an entity with managerial and governance mandates. Within this component of SMID, the WUA was expected to improve its capacities to manage water distribution, its financial operations, and resolve water-related conflicts. For the purposes of both, social mobilization and institutional strengthening of the WUA, the SMID approach prescribed a selection of so-called "social mobilizers,"' that is, a widely accepted term for teams which conduct social mobilization [20]. The social mobilizers were responsible not only for the dissemination of the information about the role and usefulness of the WUA to the various stakeholders as mentioned above, but also (and with prior training) for the formation of subclusters identified as the Water User Groups (WUG). Formally, WUG were defined as autonomous informal self-organized groups of people united by the proximity of their land to a particular irrigation source, that is, canal/ditch/pump (later called a "hydrological unit") who manage their own irrigation system to support WUA and account to it [22]. WUG, thus, represented a lower level in a multi-tier system of WUA, where the representative of each WUG participated in the decision-making by becoming a constituent in a WUA council.

4. Local irrigation system

The BMBF-UNESCO project was implemented in Khorezm province, 1 of the 12 provinces of Uzbekistan, which adjoins the environmental damaged Aral Sea and where about third of population lives below the poverty line of 1 USD per day [16]. Located 250 km south of the present shores of the Aral Sea, it covers 6800 km^2 of dry arid desert of which 270,000 hectares are used for irrigated agriculture [16]. The climate is arid with hot and dry summers and cold winters with precipitation of less than 100 mm per annum [16]. Irrigated agriculture is the mainstay of economy in the province accounting for about 67% of the total regional GDP [17]. Of 1.5 million of Khorezmian population, over 70% live in rural areas engaged in cotton, wheat, and rice production as private farmers or peasants [4]. Private farmers crop cotton, wheat, rice, and fodder maize [4]. Cotton occupies 50% of irrigated cropland and consumes about 40% of the total water supply of the region [4]. It contributes 16% to the GDP and earns almost all of the total export revenues of Khorezm province [4]. As explained above, the production of cotton and wheat follows the state procurement system, that is, the government enforces regulations on the acreage for each crop and production quantities to be submitted to the state at the fixed price, also determined by the state. In return, it ensures supply and delivery of water, diesel, fertilizers, and some other required inputs [4]. All this applies to private farmers only. Small holders cultivate potatoes, vegetables, fruits, as well as wheat and fodder [18]. They have garden plots around their houses typically about 0.12 ha and an additional plot of land of 0.13 ha called "tamorka" [18]. These tamorka plots comprise about 20% of the irrigated land of Khorezm and play significant role for the livelihoods of the households [18]. Crops in Khorezm are cultivated with a peculiar rural ecology due to high soil salinity annual leeching of the fields, and extensive irrigation are fundamental necessities [18].

Based on such a system of pre-determined production of strategic crops, irrigation management entities at national, regional, and local levels determine crop water requirements and develop delivery plans for each cropping season. Khorezm is located in the tail end of Amu Darya River and relies on upstream areas for water supply. Its irrigation infrastructure consists of about 5 km of water diverted from Amu Darya River through 16,000 km of irrigation channels [18]. During droughts, the water is reduced by 40% [18]. The arriving water is partly stored in a local Tuyamuyun water reservoir, and its volume is then rationed. Irrigation is a key factor for the fulfillment of production quotas. Quantity of water to be allocated is determined on the basis of the size of the irrigated areas, types of crops, and the irrigation norms determined by the state [6]. The allocation of water supply is carried out by the Ministry of Agriculture and Water Resources and passed on to basin irrigation system authority, to sub-basin irrigation authority to the Water User Association (WUA) [19].

The technical delivery of irrigation water through maintaining and operating the irrigation and drainage network is the responsibility of the state water management organizations such as the Main Canal Management units of the sub-basin irrigation system authority. But they are known to be unable to provide these services due to inadequate, human, financial, and technical resources [17]. Similar criticism applies to on-village operation of WUAs which are formally responsible for maintenance and operation of irrigation infrastructures within their areas [17].

5. Methodology

The aim of the research was to explore the everyday practices of the local women smallholders as their agriculture was being transformed toward sustainable practices. Fieldwork took place in Spring and Summer, 2011 in Khorezm province in Uzbekistan when the BMBF-UNESCO project was nearing its end. Ethnographic approach was selected for this study to capture and document the nuanced and complex nature of the everyday lives of the informants as immersed in social practices, institutional structures, and a local culture. Participant observations and in-depth interviews were used with individual women smallholders and members of their families. A total of 40 local women smallholders provided information in the in-depth interview and also allowed the researcher to conduct participant observations in their homes, fields, gardens, etc. All these women had kitchen gardens and tamorka where they cultivated and all of them had their male partners away from home in labor migration.

A total of nine key informant interviews were carried out with farm managers, representatives of WUA, and upper institutions of water management. Expert interviews were also carried out in Bonn, Germany, with the implementers of the BMBF-UNESCO project. Analysis of institutional text was also used.

6. Women's costs of sustainability

It is in this context that I find it important to describe the everyday struggle women smallholders in Khorezm live through as they ensure the livelihoods and subsistence for their families. These women typically cultivate a backyard garden and an additional plot of land located in some distance away from home. The backyard gardens are used intensively for growing vegetables and fruits. Hey till the land by hand with shovel and hoe. Double cropping is widely used to ensure harvest of potatoes and onions in the beginning of the agricultural season and late cropping of beans, carrots, maize, sorghum, and millet. The tamorka plots are used twice each season for producing winter wheat followed by rice or maize in the summer. Most household also keep animals and poultry. About 50% of the stallholders also work during the agricultural season on the private farmers' land for cash or in kind payment [23]. Household food production and agriculture are essential for the food and livelihood security for most rural household despite other income generating activities that the household members can become involved in [2].

The household labor is divided according to gender and generations [24]. Women are typically responsible for keeping the house in order, gardening the household plots and kitchen gardens, weeding, milking cows, processing food, and carrying out small-scale trade [25]. Women make up a large proportion of the sub-contracted workers in the private farmers' fields. Men, if they are not abroad seeking work, are normally responsible for arranging agricultural contracts, arranging irrigation turns, and irrigating the household plots. Children from age of 10 work alongside adults in the fields; at even younger age, they herd animals and help with gardening, food processing, and house chores. Elderly people often look after very young children, and their pensions provide extra cash income.

Labor migration has become an important source of income in Uzbekistan, and like in many post-soviet countries, rural households rely heavily on remittances for their cash income. Between 2000 and 2014, the total number of labor emigrants from Uzbekistan varied between 600,000 and 700,000 persons of which about 550,000 migrated to Russia [26]. Remittances from Russia only form 16% of Uzbekistan's economy [26]. Because migration is predominantly undertaken by males, women's workload has greatly increased [2]. Most labor migrants tend to be largely absent during the agricultural season leaving this burden entirely on the shoulders of those who stay at home. Women acquired new tasks such as soil fertilization, planting, irrigating and harvesting, as well as learning to organize their time to accomplish their intensified work. Ethnographic observations of smallholder women's everyday lives demonstrate their packed schedules which begin from dawn and last till midnight with only a short break in the extreme heat of the middays. The daily work includes cleaning the outside area, tending to animals and poultry, cultivating their fields, cooking meals, producing dairies, baking bread from scratch, doing laundry, harvesting vegetables or fruits, working in the garden, milking the cows, cleaning around the house, etc. Days become so busy for these women that sometimes ethnographic observations could not include conversing with them due to her attention labor-intensive tasks, noise, voices of crying or playing children around. All these activities are performed with little or no basic household equipment, running water,

http://dx.doi.org/10.5772/intechopen.79732

or piped gas. For example, baking bread is done outside with the use of mud stoves heated by firewood that women must prepare in advance. This extends these women's labor inputs by large margins. Food security is maintained using various means including producing sufficient supplies of canned vegetables and fruits which women regularly do in the summer. Canning is a good example of the complexity of their everyday work. Observing one of the respondents, Nargiza, does it demonstrates that it involves an entire day of concentration, damage control, and coordination. Nargiza woke up earlier that morning to make sure that she does the cleaning and milking of the cow before her canning endeavor. She brought buckets of water from the community well and used it to wash about 30 big glass jars which were then sterilized with a use of an old boiling kettle. Each jar was put on the top of it upside-down and boiled for about 5 min. Lids were sterilized, too, in a separate kettle. At the same time, she washed cucumbers, onions, and garlic and cleaned them of endings. Then she washed tomatoes, chopped some of them and whirled the pieces in an old semi-automatic washing machine, and rubbed them through a sieve. The resulting tomato juice was then boiled in of the three large pots built-in the mud stoves outside the house. In another pot, she would boil the vegetables in water. She would then bring a hot sterilized glass jar from the house and fill it with boiling vegetables. For this she would use a ladle and fish the vegetables from it with her bare fingers. The jar will then be filled with boiling tomato juice. She would then put salt and vinegar and put the lid on top of the jar for further tightening. This work took place at 45°C heat and interrupted by occasionally feeding the oven with brushwood, bringing clean water and taking away the dirty one, and attending to small children to prevent them from harm.

This illustration is useful in understanding the reality of smallholder women's everyday routine work. Not to forget that cattle breeding and cropping, shared with male partners, are now completely done by women themselves. The double burden makes the lives of these women dense, busy, and hectic, even though they do not complain but see as something that simply "must be done." Most smallholder households grow in their field food that provides almost full subsistence for their families for at least 10 months. For example, tamorka fields normally yields about 1 ton of flour which is sufficient for 12 months for a family of 8 people. Families which planted potatoes after harvesting wheat would have enough to consume it throughout a year.

However, there are risks and serious challenges to women's successful small-scale agriculture, which may undermine wellbeing of family. I learn about these challenges as I talked to women who complained that they had very little opportunities to irrigate their fields. The irrigation water in the village is rationed and arrives once in 2 weeks and women often miss it. Nargiza, for example, said that she was away from home and did not "catch" the water twice in the season. Other respondents shared experience such as Munara's who "must open her ditch upon hearing about the water arrival. The water can arrive at any moment during a day or night. If a person is not at home, the water bypasses this person's land." The problem goes beyond women's not having consistent and reliable information. Some of them complaint that "even if we know that the water is there, the water is limited and there is no guarantee that it will reach us …" Another respondent shared that last time the irrigation water arrived to their village, they "did not manage to irrigate their kitchen garden and field because after the private farmer had used the water, nothing was left for them." Such accounts clearly demonstrate that women-smallholders experience difficulties with accessing the irrigation water and suffer a great degree of uncertainty about not only "when" but also about "whether" they would be able to irrigate their fields. This uncertainty worries them because failure to access the irrigation endangers the success of their agriculture. These women learnt to use their specific knowledge and engage in various strategies to ensure that they do irrigate their plots. One of the respondents from a tail end part of her village shared her strategy, "If I see that ilatkom [a member of a village council] is going to the village council leader I know they will discuss water. So, I wait till he goes back and then run to him and ask about when the water can be expected." Another respondent said, "If I see on the street a hydro-technician, I run to him and ask when the water will come." Another interview demonstrated even more ingenuity, as she told me: "I know that the water will come soon when I hear the gritting sound coming from the farmer's land. I know this is his pump being started. Then I know there will be water."

These methods, simplistic as they appear, are, in fact, hard work, too. The women must physically and regularly watch for the mobility of the individuals, stay alert to "catch" them as they move around the village. These methods require that women develop and maintain good relationships with these few individuals whom they turn for information. They are then required to engage in small talk, display friendliness, deference, empathy, concerns, etc., while simultaneously suppressing their own feelings, frustrations, and anticipations. Literature classifies it as emotional labor and describes the emotional labor economy is an unfair and stressful work factors associated with negative attitudes, behaviors, and poor health [27]. Women smallholders must maintain their everyday agriculture in the conditions of high uncertainty. When the water for irrigation will arrive, for how long and how much are the questions that are often left with no clear and systematic responses. Living with such level of uncertainty is also a hard psychological work which involves anxiety, worry, out of control, hopelessness, and helplessness. These women must learn to live part of their lives in the conditions of chaos and randomness which can be very scary.

Finding a way to manage and live with such uncertainty in order to bring a sense of order and predictability. But the reduction of uncertainty also involves considerable amounts of physical work. Uncertainty forces the women smallholders to resort to a number of time-consuming and labor-intensive strategies. Most women must physically go to canal to see if the water is flowing. The distance to the canal may range from 50 m to more than several kilometers of unpaved roads from a woman's house. For instance, Firuza takes 2 h by her donkey-harnessed cart to reach her field and look at the canal. If the water is not flowing, this long journey is undertaken in vain. If the water is there, she queues with other smallholders and waits until she can open her ditch and let the water flow into her plot of land. Depending on the water pressure, irrigating one plot takes from 40 min to 5 h. This adds up to long hours of work, added to the additional hours of journey back and forth to the village. Mavluda walks or uses her bicycle to go to the canal. By bicycle, it takes her 20 min to reach the place, and she has to do this once in every 2–3 days during the vegetation season. She says: "There is no one to ask or to telephone. Once I was lucky and learnt about the water from a neighbor who is employed at the farm and knew about it." However, regardless of the creativity, they introduce into their already multilayered and complex everyday work, they often fail to do the irrigation work because they either do not get timely information or do not manage to be physically present in their fields when the water comes, or else the water is already used up.

http://dx.doi.org/10.5772/intechopen.79732

These outcomes, apparently, contradict the original policy/project promise and their effects in relation to these smallholders. The project's approach and participatory promise were to bring social assistance to the most vulnerable groups and, as mentioned above to "improve livelihoods of the rural inhabitants." Elsewhere, I explored in detail these contradictory findings and map out institutional process, which organized the local experiences in such a way [28]. Here, the argument focuses on making visible the voices of these women smallholders who were made invisible in how national and international address the Aral Sea crisis through bringing the concept of sustainability into these people's everyday lives. The national effort to make irrigation water use more efficient introduced the policy of rationing which was carefully regulated through established hierarchies, procedures, operations, etc. The international European actors engaged in transfer of knowledge and expertise and worked with the local WUA. Yet, they all missed not only the needs and interests of the women smallholders but also their potential contributions.

Ethnographic data show some presence of women smallholder organization within the village. To provide just one example, it is useful to turn to a smallholder woman whom I call Gulnara. Gulnara is a retired school teacher whose neighbors were refused irrigation services by the WUA due to a long history of fights between the WUA and the people. Gulnara took on a mediating role and served her street for the last 5 years. When interviewed, she supervised irrigation of about 50 households and managed the organization of the related processes through village level lobbying, mobilizing people to clean the irrigation infrastructure, collecting fees, and keeping careful accounting of her work. Stories like Gulnaras suggest that local women engage in social activism and actively engage in the kinds of local dynamics that the project aimed at attaining. However, neither Gulnara nor other women like her have never been invited to any project activities and remained unknown to the project staff. I tend to see this loss for the project's commitment to a bottom-up approach and its goal to bring more social justice.

7. Conclusion

Ethnographic data showed that women smallholders have specific needs and interests in having reliable and sufficient access to water to continue growing their crops. This was vital to their subsistence, livelihoods, health, and lives. These women also contribute to community-based local water management leading water distribution and taking control of shared resources. However, these active women experience hardships in obtaining dependable access to irrigation sources. In the context of lacking any systematic information about scheduling of irrigation water, its delivery and quantity, these women engage in a number of creative strategies to learn about water availability. However, these strategies involve considerable amounts of physical work as well as emotional labor that must be invested in exchange of valuable knowledge about the water. This happens in the context where most of the women's already busy workloads have been added considerably due to their male partner out migration. Importantly, this happens in the conditions of the government's and international project's policies to accord water management and water use with the notion of sustainability. The government politics of rationing was introduced as part of sustainable and rational water use. The BMBF-UNSECO project's enhancement of WUAs was another promising social innovation which was expected to lead to improvements for all. However, as observed from the data on the ground, the trickle-down effect did not happen for the women smallholders. They continue to suffer shortages of irrigation water, risking their own subsistence, and those of their families, while the high-level talks about sustainability continue to overwhelm various international fora. At the moment, as what the women's experiences show, sustainability is achieved at the expenses of women smallholders' time, health and, ultimately, lives. These outcomes of water sustainability as a policy and practice are unfair and contradictory, thus, call for attention and subsequent actions from professionals, developers, planners, and policy makers. What appears necessary at this point is to promote policy and development action that would base their strategies on broad and in-depth research of the everyday relevancies and actualities among the prospective beneficiaries in order to be able to initiate discussions about how to integrate their interests and concerns into sustainability programming. Serious attention to how women can both benefit from and contribute to water sustainability policies and projects must become habitual in the development and professional circles.

Conflict of interest

No conflict of interest is involved in this publication and related research.

Notes/Thanks/Other declarations

The research was made possible with funds from the German Academic Exchange Program (DAAD) and Fiat Panis Foundation.

Author details

Elena Kim

Address all correspondence to: kim_el@auca.kg

American University of Central Asia, Bishkek, Kyrgyzstan

References

[1] Ismailova A, Baynazarov E. Analysis of the agrarian land reform in Uzbekistan during the Soviet era and after transition. EU Agrarian Law. 2015;**2**:61-67

[2] Kandiyoti D. The cry for land: Agrarian reform, gender and land rights in Uzbekistan. Journal of Agrarian Change. 2003;**3**:225-256

[3] Micklin P. The Aral Sea disaster. Annual Review of Earth and Planetary Sciences. 2007;**35**:47-72

[4] Rudenko I. Value chains for rural and regional development: The case of cotton, wheat, fruit, and vegetable value chains in the lower reaches of the Amu Darya River, Uzbekistan [thesis]. Center for Development Research: University of Bonn; 2008

[5] Don Van Atta. Land and water politics in Uzbekistan. The National Council for Soviet and East European research. Available from: https://www.ucis.pitt.edu/nceeer/0000-000-00-VanAtta.pdf

[6] Djanibekov N. A microeconomics analysis of farm restructuring in the Khorezm region, Uzbekistan [thesis]. Center for Development Research: University of Bonn; 2008

[7] Spoor M, Visser O. The state of agrarian reform in the former Soviet Union. Europe-Asia Studies. 2001;**53**:885-901

[8] Humphrey C. Marx Went Away. But Karl Stayed Behind. Ann Arbor: University of Michigan Press; 1998

[9] Uzbekistan Economy Profile [Internet]. 2018. Available from: https://www.indexmundi.com/uzbekistan/economy_profile.html [Accessed: 15-06-2018]

[10] The Permanent Mission of the Republic of Uzbekistan to the United Nations [Internet]. Problems of the Aral Sea and water resources of Central Asia. Available from: https://www.un.int/uzbekistan/news/problems-aral-sea-and-water-resources-central-asia [Accessed: 15-06-2018]

[11] Panda A. How the soviet union created Central Asia's worst environmental disaster. The Diplomat. 2014. Available from: https://thediplomat.com/2014/10/how-the-soviet-union-created-central-asias-worst-environmental-disaster/ [Accessed: 15-06-2018]

[12] Worsnip P. U.N.'s Ban urges Central Asia talks on shrinking Aral Sea. Available from: https://www.reuters.com/article/us-uzbekistan-un-sea/u-n-s-ban-urges-central-asia-talks-on-shrinking-aral-sea-idUSTRE6330QM20100405 [Accessed: 15-06-2018]

[13] Guterres A. Statement following his visit to the Aral Sea. 10 June 2017. Available from: https://www.un.org/sg/en/content/sg/speeches/2017-06-10/secretary-general's-aral-sea-statement [Accessed: 15-06-2018]

[14] Vlek P, Martius C, Schoeller-Schletter A, Lamers J. Economic Restructuring of Land and Water Use in the Region Khorezm (Uzbekistan) (Project Proposal for Phase I). Bonn: ZEF Working Paper Series; 2011

[15] Zander U, Moll P. Transdisciplinary research: Not an easy exercise. ZEF News. 2008;**20**:1-3

[16] Mueller M. Cotton, agriculture and the Uzbek government. In: Wehrheim P, Schoeller-Schletter A, Martius C, editors. Continuity and Change: Land and Water Use Reforms in Rural Uzbekistan. Socio-Economic and Legal Analyses for the Region Khorezm. Halle: IAMO; 2008. pp. 89-105

[17] Manschadi AM, Oberkircher L, Tischbein B, Conrad C, Hornidge A-K, Bhaduri A, Schorcht G, Lamers JPL, Vlek PLG. "White gold" and Aral Sea disaster—Towards more efficient use of water resource in the Khorezm region, Uzbekistan. Lohmann Information. 2010;**45**:34-47

[18] Bobojonov I. Modeling crop and water allocation under uncertainty in irrigated agriculture. A case study on the Khorezm region, Uzbekistan [thesis]. Center for Development Research: University of Bonn; 2008

[19] Martius C, Froebrich J, Nuppenau E-A. Water resource management for improving environmental security and rural livelihoods in the irrigated Amu Darya lowlands. In: Brauch HG, Oswald Spring Ú, Grin J, Mesjasz C, Kameri-Mbote P, Chadha Behera N, Chourou B, Krummenacher H, editors. Facing Global Environmental Change: Environmental, Human, Energy, Food, Health and Water Security Concepts 4. Berlin/Heidelberg/New York: Springer; 2009. pp. 749-762

[20] Abdullaev I, Franz J, Oberkircher L, Hoffman I, Nizamedinkhodjaeva N, Ataev J, Ul Hassan M, Lamers J, Tischbein B, Schorcht G, Jumaniyazova Q, Djanibekov N. Work Plan of Follow the Innovation Activity. Group 3. Improving Livelihoods of Rural Inhabitants through Better Water Management. Improving WUA Performance through Social Mobilization and Institutional Development. Bonn: Center for Development Research, University of Bonn; 2008

[21] Ul Hassan M, Hornidge AK. 'Follow the Innovation'—The Second Year of a Joint Experimentation and Learning Approach to Transdisciplinary Research in Uzbekistan. Bonn: ZEF Working Paper Series; 2010

[22] Abdullaev A, Kazbekov J, Manthritilake H, Jumaboev K. Water user groups in Central Asia: Emerging form of collective action in irrigation water management. Water Resources Management. 2009;**24**:10-29. DOI: 10.1007/s11269-009-9484-4

[23] Veldwisch GJA, Bock BB. Dehkans, diversification and dependencies: Rural transformation in post-Soviet Uzbekistan. Journal of Agrarian Change. 2011;**11**:581-197

[24] Nizamedinkhodjayeva N, Bock B, Mollinga PP. The role of subsistence agriculture for rural livelihoods in the Khorezm province of Uzbekistan. In: Manschadi A, Lamers JPA, editors. Restructuring Land- and Water Use in Khorezm the Region, Uzbekistan. Bonn: Centre for Development Research (ZEF), University of Bonn; 2013

[25] Spoor M. The rural development challenge of transition. In: Spoor M, editor. The Political Economy of Rural Livelihoods in Transition Economies; Land, Peasants and Rural Poverty in Transition. Abingdon: Routledge; 2009. pp. 1-10

[26] Prague Process. Republic of Uzbekistan. Migration Profile Light. Available from: https://www.pragueprocess.eu/en/documents/category/5-migration-profiles

[27] Jeung D-Y, Kim C, Change S-J. Emotional labor and burnout: A review of the literature. Yonsei Medical Journal. 2018;**59**:187-193. DOI: 10.3349/ymj.2018.59.2.187

[28] Kim E. International development and research in Central Asia. Exploring the knowledge-based social organization of gender [thesis]. Center for Development Research: University of Bonn; 2014

Chapter 6

Comparison of Water Resources Community Self-Management Mode between China and Tanzania

Dan Li and Mngereza Mzee Miraj

Additional information is available at the end of the chapter

http://dx.doi.org/10.5772/intechopen.84194

Abstract

Due to limited rainfall and uneven spatial and temporal distribution of water resources, water has become a restraining factor in agriculture and livestock production of China and Tanzania. As it is most considered as common-pool resource, the management of water resources is a complex issue in agricultural and pastoral industry. Traditional water management modes include nationalization and marketization, but complete market-oriented or government management could not reach the sustainable use of water resource due to nonexclusive and interconnected features of water. Therefore, China and Tanzania introduced water resources community self-management in rural arid areas. Farmers as resource users in community conducted mutual supervision and mutual benefit to realize reasonable, fair, and sustainable use of water resources. However, community self-management is restricted by formal institution from the government of China, and Tanzania's community self-management relies on the financial and technical support from foreign NGOs; the communities' ability to obtain benefit needed to be improved. We compare water resources community self-management mode in China and Tanzania through case studies, put forward the differences of self-management mode in two countries, and analyze the characteristics of successful water resources community self-management mode.

Keywords: water resources use, community self-management, farmers' livelihood, China, Tanzania

1. Introduction

1.1. Water resources in arid regions in China and Africa

Water is the basis for the survival and sustainable development of the human society. With the development of the society and economy, the water crisis caused by the lack of water resources and water pollution has become one of the key factors restricting the economic and social development. Nowadays, arid and semiarid regions in the world account for about 40% of the total land area, while the freshwater resources on the earth only account for 1.6% of the water resources on the earth's surface. About 40% of the global population in more than 80 countries is facing serious water crisis [1]. With the trend of global warming, the area of arid and semiarid regions will accelerate its expansion, which is expected to account for more than 50% of the global land surface by the end of the twenty-first century. In addition, three-quarters of the arid and semiarid region expansion will occur in developing countries, exposing developing countries to the risk of further land degradation and aggravating the poverty of people in arid and semiarid regions [2]. The United Nations Water Conference pointed out that the next crisis after the oil crisis is the water crisis [3].

The inland arid zone of Northwest China is located at the north of the 35°N and the west of 106°E, including Xinjiang, the Hexi Corridor in Gansu province, and Inner Mongolia region, west of Helan mountain, accounts for about 24.5% of the total land area of China [4]. The northwest arid region consists of mountains and basins. Rivers originate from the mountain area flows to the basin. The distribution of water resources determines that the surface runoff and groundwater resources of the area are the key factors and ties to maintain the economic development and ecological environment balance of the middle and lower reaches. The climatic conditions of inland drainage area show a significantly difference, with precipitation ranging from 300–1000 mm in the alpine region to 100–200 mm in the plain region, and seasonal variation was obvious. Precipitation is mostly concentrates from June to August, and drought was common in winter and spring (**Figure 1** [5]). Under the constraints of water resources distribution and climatic condition, the inland river valley ecological environment system usually forms the argo-pastoral transitional zone with pastoralism and irrigated agriculture, which is the fragile ecological environment zone. With the increase of population, the development of social economy, and the exploitation and utilization of soil and water resources, a series of hydrological and ecological environment changes have been occurred. As a result of the drought caused by the reduction of water resources, grassland reclamation, and overgrazing, the grassland area in river valley is reduced and seriously degraded. Grassland degradation caused the decline and disappearance of some dominant herbage species and thus the decline of biodiversity [6]. From 1958 to 2005, the forage yield in Northwest China decreased by 75.4% [7].

In sub-Saharan Africa (SSA), drought area accounts for 20% of land but accounts for over 80% of the affected population [8]. Much of the continent is dependent on rain-fed agriculture, which makes it particularly susceptible to climate variability. Almost 70% of the labor force is engaged in agricultural work, and agriculture contributes to about 25% of average gross domestic product (GDP) across the continent [9]. The limited water resources have direct

http://dx.doi.org/10.5772/intechopen.84194

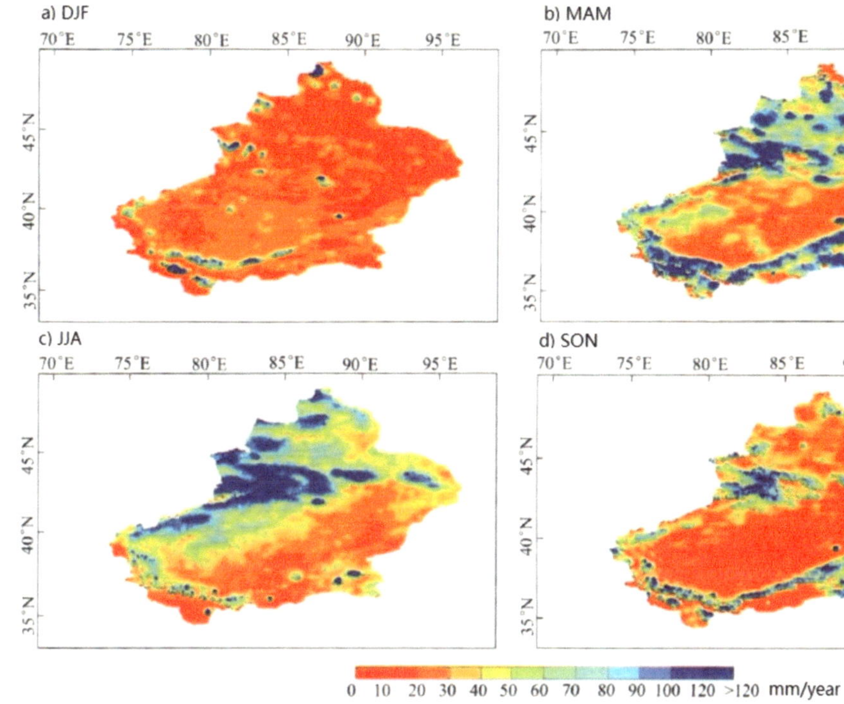

Figure 1. Seasonal mean precipitation (1950–2008) in Xinjiang, China, for (a) December–February, (b) March–May, (c) June–August, and (d) September–November.

impacts on agriculture and livestock grazing over large space and time scales. The impacts are driven by the high vulnerability of the natural environment and are exacerbated by prevailing local and external economic and political conditions [10], which can be associated with development of famine and may be accompanied by the spread of disease. The population of SSA is over 870 million people and is expected to at least double by the mid-twenty-first century. Coupled with expected overall drying with climate change, in particular in Southern Africa and parts of West Africa [11–13], there are worrisome implications for water resources sustainability use and food security.

Water resources in SSA is linked to the high seasonal and inter annual variability in rainfall (**Figure 2** [14]). In general, seasonal rainfall higher than 500 mm is required to sustain healthy agriculture, highlighting the tenuous nature of agro-pastoral livelihoods in the transitional regions between semiarid and arid regions in some parts of SSA. In northern Tanzania, the rainy season is generally from November to April, and well-defined dry season is in July–September.

The shortage and the imbalance of spatial and temporal distribution of water resources have become the bottleneck of economic growth and social development in arid regions. With the increase of population and the expansion of industrial and agricultural production and urbanization, the residents in arid regions have an increasing demand for water resources. Many ways have been taken to expand the scale of water resources development. While obtaining

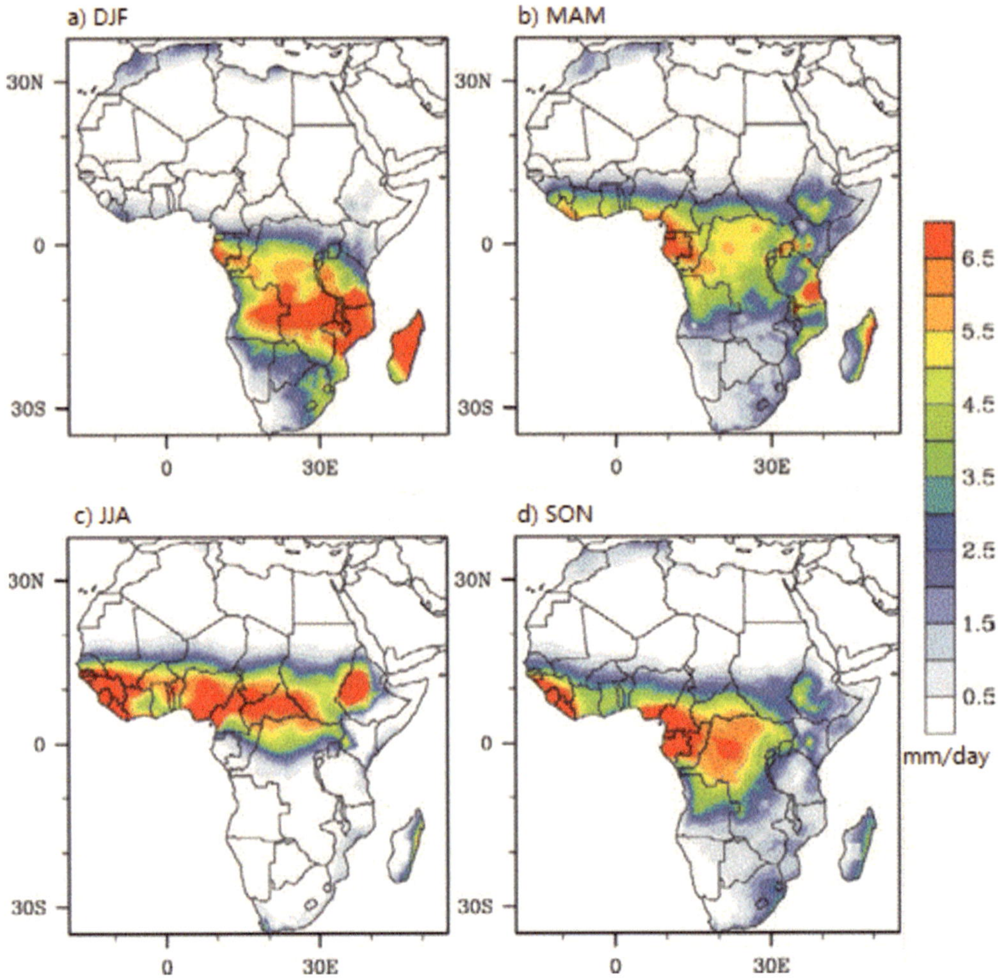

Figure 2. Seasonal mean precipitation (1950–2008) in Africa for (a) December–February, (b) March–May, (c) June–August, and (d) September–November.

temporary economic benefits, it has caused serious negative impacts on the ecological environment. Understanding the ecological environment and the farmers' livelihood needs and changes in arid regions; managing limited water resources scientifically and rationally developing the maximum economic, social, and ecological benefits of scarce water resources; and ensuring the sustainable use of water resources have always been the key concerns of the arid region research.

1.2. Management of water resources in arid regions

Water resources management can be divided into government regulation, water rights trading market, and community self-organization management according to the different distribution modes of water resources property rights.

Due to the mobility, recycling, and public characteristics of water resources, most national laws stipulate that the ownership of water resources belongs to the state, and the state has the right of allocation and final decision-making of water resources. In fact, the ownership, management, and use of water resources are separate. The ownership of water resources in China belongs to the government, the use of water resources by industries and agriculture should be under administrative permission, and the water use license cannot be traded on the market [15]. The government allocates limited water amount by administrative means, and researchers explore various methods to optimize water allocation, in order to maximize the economic, social, and environmental benefits. However, due to the conflict of interests among the water use stakeholders, it is impossible to achieve the optimal allocation of water resources in practice. At the same time, the regulation of water price fails to reflect water value, and this rationing system eventually leads to the general expansion of demand, further exacerbating the contradictions between the stakeholders and increasing the difficulty and confusion of management.

On the other hand, developed countries usually adopt the method of establishing water resources trading market. Based on the clear definition and initial use right of distribution of the water resources, the use right of water resources exchanges among regions, basins, upstream and downstream, industries, and households through market mechanism. Under the law of value, water price and water resources value could be adjusted and matched, and the distorted situation that water resources are priceless or low could be changed. According to the development of the market economy, economic leverage is used to regulate water prices, and the government only carries out macro supervision in this process [16]. However, as water is an irreplaceable vital resource, complete marketization will also face many problems according to water resources characteristics, such as the definition of initial water resources allocation mode and water users' short-sighted behaviors driven by interests, which will cause the failure of water resources market.

Besides government and market, Ostrom proposed the third option in "Governing the Commons," namely the self-organization and management of common-pool resources [17]. Based on the analysis of several classic models including Hardin's "The Tragedy of Commons" and "Prisoner's Dilemma" and Olson's "The Logic of Collective Action," the conflict between individual rationality and collective rationality was drawn. According to Ostrom, the defects of traditional game analysis methods and the theoretical assumptions they rely on were deviated from the real situation, such as rational person assumption, complete information access, independent action, noncommunication, and first-order game. Ostrom took small-scale common-pool resources as an example and demonstrated that a group of limited rational person communicated and interacted with each other in the process of sharing natural resources. They could obtain more information on resources and other actors' behavior and develop effective common-pool resources use contract through self-raised funds. Ostrom analyzed the possibility of community self-management theoretically. In Ostrom's theory, although limited rational actor did not have complete information, they could increase their understanding of other actors through communication in the process of the game, fully understand each person's influence on common-pool resources, and then change their own strategies to obtain more benefits.

In Ostrom's Institutional Analysis and Development Framework (IAD), collective action of resources needed to solve three problems, such as the problem of supply of the current institution, the problem of credible commitment, and the problem of mutual monitoring. As for the supply of the institution, Ostrom believed that cooperation balance should be generated through multiple games among community members based on current institution, in order to form a series of mutual beneficial situation and an informal system of community mutual trust. As for credible commitment, Ostrom argued that self-management groups should develop effective regulations and take appropriate supervision and sanction measures to ensure that community members follow the rules. As for mutual monitoring, Ostrom believed that after the establishment of regulations and the commitment to follow the rules, the implementation of the regulation and the allocation and use of common-pool resources in accordance with regulation should be monitored.

Therefore, we compared two typical cases of water resources community self-management in China and Tanzania; described the details of the cases from the supply, credible commitment, and mutual monitoring aspects; analyzed the internal difference between China case and Tanzania case; and thus put forward effective community self-management mode that has a positive impact on natural resources and the livelihood of farmers.

2. Material and methods

2.1. Study area

2.1.1. China

Xinjiang's Yili valley agro-pastoral zone is stratified by elevation, transitioning from low-altitude semiarid agriculture at elevations below 1000 m to humid alpine meadow pastoralism at elevations above 1000 m. The annual precipitation below 1000 m is 400–500 mm [18]. With relatively abundant snowmelt from the Tian Shan mountains, the valley's lowlands and riparian corridors provide a significant share of Xinjiang's irrigated agriculture, whereas the middle and upper regions of the mountains are humid alpine meadow grassland that has been used for extensive livestock grazing (mainly sheep but also cows, goats, horses, and some camels) for a thousand years.

M village is located on the western slope of the Tian Shan mountains in the headwaters of the Yili River, in the Yili Kazak Autonomous Prefecture. Pastoralism and agriculture coexist, and the former plays a dominant role. There were 558 households with 2273 people, of which 50% were Kazakhs (village statistics). Natural pasture area is about 9333 ha. Farmland area is about 504 ha.

2.1.2. Tanzania

The study was conducted in Hai district specifically at Saaki spring as a case study. In the recent years, there has been a tendency of cutting trees around the Mountain Kilimanjaro on the side of Hai district which impact in the shortage of water around the district causing serious problem at Saaki spring and Hai district as a whole. Hai district which is situated in the

Northern part of Tanzania is among the six districts forming Kilimanjaro Region. The district is subdivided into three divisions which are Lyamungo, Machame, and Masama. The district has 14 wards, 60 villages, and 11 urban streets. Saaki spring is the biggest source of water which serves people who live in Hai town where the district headquarters is situated and is also serving people who live in the villages. Generally, the Saaki spring is approximated to serve the population of more than 58,003 people who live in the villages and streets.

S village is located in the middle of the district, which belongs to Masama division. The list of water user households was gathered in village registers. Most of the villagers participated in agriculture production. The population of the village was 3793 in 532 households.

2.2. Data and methods

The study used a qualitative approach to describe the current status of water resources community self-management in China and Tanzania case. As supplementary, quantitative approaches help to measure data from the field work study. The two approaches complemented each other in gathering data to create valuable information for understanding community participation in water resources management.

Primary data on community participation in water resources management were collected from the respondents. Field research was completed using semi-structured interviews with households in 2015. Interviewees were selected by purposive and simple random sampling. Eighty-three households in M village (China) and 80 households in S village (Tanzania) were interviewed, more than 15% of the total household number in two villages. The purposive sampling technique was used to select the key informants from the village level who were knowledgeable and responsible for developmental issues and water resources management in their respective areas of work. Simple random sampling technique was used to select households in the study area to represent the specific and detailed information. Interviews focus on water use and management in agriculture and livestock production and the perceptions and opinions of interviewees on environmental and social changes. Additional interviews of local government officials, water engineers, and NGO technicians provided overall information.

3. Results

3.1. Institution supply: use rights of water resources

3.1.1. China

In the late 1970s, as China transitioned from a planned economy to a market economy, the Household Land Contract System (HLCS) was implemented. The land was contracted to individual households while formally remaining the collective ownership. According to the HLCS, all agricultural outputs are owned by the household except for the state agricultural tax (which was canceled in 2006). Land use privatization greatly increased labor productivity and rural economic development and thus helped numerous farmers climb out of poverty (Lin, 1994).

However, water resources have the integral characteristics. The law claims that water resources are owned by the state. The Chinese government has many departments involved in water resources management, but there is no independent and complete water resources management center, which lacks unified and coordinated management at the government level. Compared with industrial water use, agricultural irrigation water is generally dispersed, random, and low marketable. It is difficult to establish a standard water resources trading market, and under the premise of state-owned water resources, it is also difficult for farmers to obtain the independent water resources use rights and to be water resources traders. On the other hand, since ancient times, the nomadic Kazak people lived in tribes, shared information with their relatives, and helped each other in agricultural and pastoral production, providing a good cultural foundation for the community self-management model.

3.1.2. Tanzania

The land in Tanzania belongs to private landowners, and landowners were entitled to spring water on their land. Before African independence, the state played a negligible role in the allocation of water rights and the development of water resources [17]. After 1964 (Republic of Tanzania foundation), the water use rights were controlled and regulated by the state, but landowners still had the right to use public water in public streams.

The right of private owners to use water in rural areas which had its source on the land or flowed over the land was a direct consequence of their landownership. Although there was no finality over the ownership of water, the use of water was derived from and linked to the ownership of land.

China and Tanzania all experienced land use privatization. However, the water property has always been rather vague. Water resources was owned by village collective in China, but owned by landowners in Tanzania. China's government has much more authority in water management than Tanzania. However, water use rights need to be distributed to private household in practice. No matter in China or Tanzania, community is the actual main body of the water resources use.

3.2. Credible commitment: water use regulation establishment

3.2.1. China

In M village, rain-fed farmlands and livestock drink water use were mostly from river and precipitation, which did not relate to the allocation of water resources. The water resources community self-management was mainly reflected in the irrigation of irrigated farmland through water canal.

The Household Land Contract System was introduced in 1984, according to the privatization use of farmland, the sorted by position and used in turn allocation way of water resources for each household was formed. Due to the unstable water volume of the canal, farmers did not pay the irrigation water fee at first, and the water resources management was quite chaotic. The upstream households of the canal might use more water, and downstream households had no enough water. Or, the imperfect rotation management mechanism led to missed watering of some households. Now, under the guidance of the village government, M village had gradually formed a mode of community self-management of water resources allocation. First of all, a water manager was elected on the villager meetings at the beginning of each year, and the farmers in village acted as water manager in turn. The water manager was responsible for collecting water fees and managing the canals, resolving farmers' water use disputes, and prohibiting water theft. The salary of water manager is 10% of the water fee. The water supply of the canal was from June to September. The water fee increased from 24.05 dollar/ha in 1984 to 72.14 dollar/ha in 2015. Each household paid 50% of the water bill in June and another 50% in September. Besides the salary of the water manager, the remaining 90% of the water fee is paid to the village cooperation for the reinforcement and seepage prevention of the canal. Households who paid water fee took turns to irrigate their farmlands from upstream to downstream, the diameter of irrigation pipes was fixed, and each household could irrigate for a maximum of 48 hours, after which irrigation water was rotated to the next household (**Figure 3**).

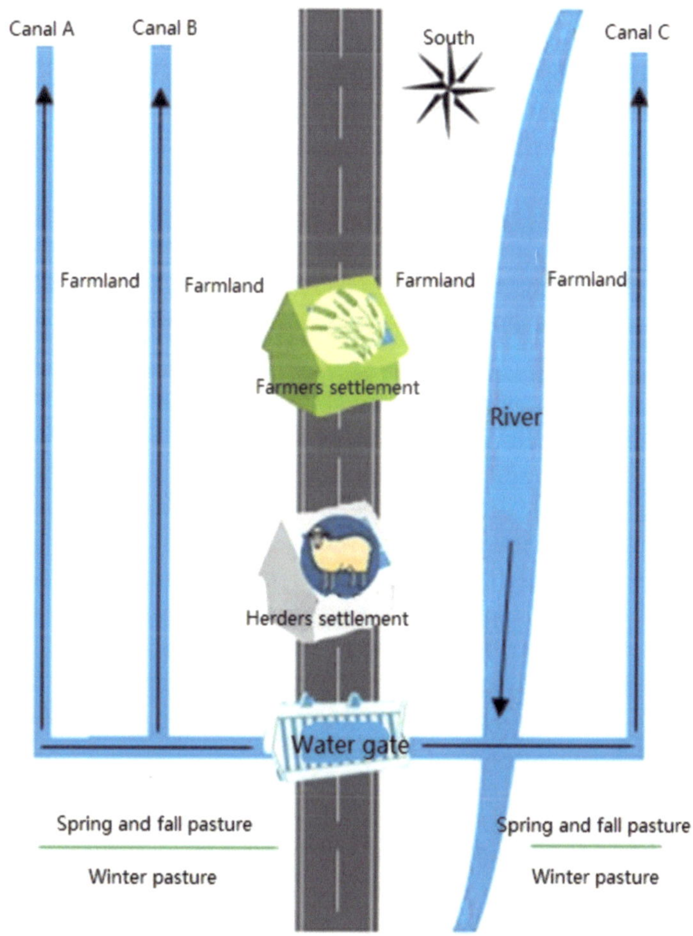

Figure 3. The canal diagram of M village.

3.2.2. Tanzania

Water resources community self-management was implemented in Tanzania by UBWS (Uroki-Bomang'ombe Water Supply). They sent technicians, made guidebook and gathered villagers' representatives to discuss the baseline environmental conditions, mapped and interpreted water resources present situation, analyzed resource user and stakeholder, and developed the action, monitoring, and evaluation plan. However, when the farmers were asked whether they knew who was responsible for the management of Saaki spring, 55% of the farmers had the idea that the one who is responsible for Saaki spring management is the water authority and district government, and only 12.5% had the view that it was managed by the community-based group.

More than 50% of the farmers claimed that they were not involved in the planning activities, but they were involved in planting trees, cleaning the water source, and securing the water source. No more than 10% of the farmers involved in and participating in water regulation discussion meetings. This shows that the community has not been adequately involved in the spring management meetings, which indicates the need to seriously address it.

Fifty-five percent of the farmers disagreed on the statement that Saaki spring was managed through information sharing. There is much to be done to improve the information sharing of community in the management of the spring. What is more is that some of the community members even did not understand the existing regulations governing the spring. About 30% of the farmers were aware that cutting trees nearby the source of water, farming around the water source, trespassing, feeding animals, dumping poisonous wastes, and washing clothes were prohibited by the laws.

3.3. Mutual monitoring performance: farmers' perception and water use efficiency

3.3.1. China

As for community self-management, after the establishment of the water resources use regulations, the effective supervision and punishment mechanism were particularly important. Eighty percent of farmers in M village believed that the existing irrigation water allocation and rotation system in village are effective in improving the water use efficiency. However, there is still a gap with the optimal efficiency, which is caused by the imperfect management system and serious waste of water. In particular, although the water manager was elected by all the villagers, he/she was also belonged to farmer households in the village. When irrigating his own farmland or the farmland of his relatives, his supervision might be ineffective and unfair. On the other hand, when the water disputes between villagers occurred, water manager had no absolute authority to judge the problem as national judicial departments and also had no right to enforce households who caused the problem to compensate for damage. The water manager was just a mediator, persuaded both sides to put down the disputation and carry out a harmonious solution. Most of the time, the disputation still destroyed social mutual trust between farmer households.

3.3.2. Tanzania

From the interview, majority (81.2%) of households were not satisfied with the management of Saaki spring, while only 18.8% were satisfied. The majority (63.9%) of people claimed that they were not satisfied because they were not involved in decision-making. Moreover, 9.2% of the farmers commented that plans of managing the spring were not well implemented and the cost of connecting water was too high. In general, farmers held the view that there was poor community management of the spring water. They asserted that they want to get more involved in meetings and planning in the future.

The big challenge which faced community participation was limited involvement in the management of the spring. Other challenges included difficulty to protect the spring, poor supervision of water source, the cutting down of trees for firewood, and cleaning the spring.

4. Discussion

From the analysis of two cases of water resources self-management in China and Tanzania, the different property rights system of water resources in China and Africa leads to the different impeller of community self-management, and the final results are also different. The impeller of China is the village government, while the impeller of Tanzania is NGO. To form an effective water resources community self-management mode in arid regions, the following points need to be noted:

1. Full participation of the community member contributes to the rational allocation of water resources.

   In Tanzania, the large community was not given the chance to participate at various stages like planning, implementation, and evaluation. Only few, especially village leaders, claimed to participate in all stages. However, every villager participated in electing a water use supervisor and agreed with the water rotational use regulation in China. People needed to be involved from the earliest stage to the upper one during the self-management procedure. Water resources managed without the participation of the community in decision-making, planning, implementation, and evaluation are often not properly maintained and hence lack sustainability (NAWAPO, 2002).

   When carrying out community self-management, the majority of the farmers should be involved in participatory meetings, participatory planning, protecting the water source, supervising the water sources, and training on water source preservation. It is the responsibility of the local government to make sure that the large community is involved in the whole process. It will lead to community participation, pollution control and information sharing, and hence the sustainability of water resources.

2. Effective information sharing is conducive to the water resources use regulation.

   Information dissemination was very crucial for the community in order to promote community participation in the process of water self-management. Adequate information sharing leads to optimal goal achievement and relationship building; hence, the effective and efficient dissemination of accurate information to the public is essential. Informing and educating those who participate in community projects could make them permanently able to defend their own interests (Abrahamsson, 1977). Thus, participation supports the integration of interests through an intensive exchange of information among concerned actors and lays the foundation for cooperation and establishment of the sense of ownership for the sustainability of the water resources.

The local government could provide important technology guideline, database, experience, and ideas that could lead to practical, relevant, achievable, and acceptable community self-management solutions.

3. The combination of formal and informal institutions is conducive to the effective mutual monitoring.

Community self-management mainly relies on community informal management system. On this basis, appropriate intervention of formal systems may be helpful to water resources management. Formal institutions could ensure community members follow the rules and punish those who violate the rules more effectively. For example, the supervision and punishment in the water resources use regulations can be raised to the formal level. There are laws to be followed in the performance of the water management regulations, and an independent monitoring organization for villagers can be set up to strengthen the intensity of supervision, punishment, and mediation. On the basis of the complete participation of all members, communication, and information sharing mechanism, the involvement of the formal system can avoid the negative influence of the farmers' social relations on the mutual supervision performance within the community under the informal system.

Author details

Dan Li* and Mngereza Mzee Miraj

*Address all correspondence to: lidan0617@live.com

Peking University, Beijing, China

References

[1] Greve P, Orlowsky B, Mueller B. Corrigendum: Global assessment of trends in wetting and drying over land. Nature Geoscience. 2014;**7**(11):848. DOI: 10.1038/ngeo2274

[2] Huang J, H Y, Guan X, et al. Accelerated dryland expansion under climate change. Nature Climate Change. 2016;**6**:166-171. DOI: 10.1038/nclimate2837

[3] Hansson K. The United Nations water conference. In: Arid Land Irrigation in Developing Countries. US: Pergamon Press. Vol. 10. 1977. pp. 439-440. DOI: 10.1016/B978-0-08-021 588-4.50066-X

[4] Zheng B, Tian Z, Wang W. Analysis of recent land usage and survey in Western China. Acta Ecologica Sinica. 2004;**5**:1078-1085. DOI: 10.3321/j.issn:1000-0933.2004.05.032

[5] Yin G, Li L, Meng X, et al. A research of precipitation trend and fluctuation in Xinjiang from 1979 to 2013. Journal of North China University of Water Resources and Electric Power. 2017;**28**:20-27. DOI: 10.3969/j.issn.1002-5634.2017.05.003

[6] Wang G, Cheng G. The utilization of water resource and its influence on eco-environment in the northwest arid area of China. Journal of Natural Resources. 1999;**14**:110-116

[7] Sun H, Li W, Xu Y. Climate-productivity of grassland and its response to climate change in Ili River Basin, Xinjiang, China. Acta Prataculturae Sinica. 2010;**6**:55-61

[8] UN/ISDR. Drought Risk Reduction Framework and Practices: Contributing to the Implementation of the Hyogo Framework for Action. United Nations Secretariat of the International Strategy for Disaster Reduction; 2009. p. 197

[9] Dixon J, Gulliver A, Gibbon D. Farming Systems and Poverty: Improving Farmers' Livelihoods in a Changing World. FAO and World Bank; 2001. p. 407

[10] Haile M. Weather patterns, food security and humanitarian response in sub-Saharan Africa. Philosophical Transactions of the Royal Society of London. 2005;**360**:2169-2182. DOI: 10.1098/rstb.2005.1746

[11] Sheffield J, Wood EF. Projected changes in drought occurrence under future global warming from multi-model, multi-scenario, IPCC AR4 simulations. Climate Dynamics. 2008;**13**:79-105. DOI: 10.1007/s00382-007-0340-z

[12] Williams AP, Funk C. A westward extension of the warm pool leads to a westward extension of the Walker circulation, drying eastern Africa. Climate Dynamics. 2011;**37**(11-12): 2417-2435. DOI: 10.1007/s00382-010-0984-y

[13] Seneviratne S. Managing the Risks of Extreme Events and Disasters to Advance Climate Change Adaptation. Cambridge University Press; 2012. pp. 109-230. ISBN: 9781139177245

[14] Sheffield J, Wood E, Chaney N. A drought monitoring and forecasting system for sub-Sahara African water resources and food security. American Meteorological Society 2014;**95**:861-882. DOI: 10.1175/BAMS-D-12-00124.1

[15] Chen Z. Innovation and property right of China's water resource. Journal of Huazhong Agricultural University. 2001;**3**:1-3. DOI: 10.3969/j.issn.1008-3456.2001.03.001

[16] Duan W, Jia F. Discussion on the mode of marketisation of water resources exploitation and employment. Shanxi Architecture. 2002;**12**:113-114. DOI: 10.3969/j.issn.1009-6825.2002.12.076

[17] Ostrom E. Governing the Commons. Cambridge University Press; 1990. ISBN: 0521405998

[18] Chen J. The Sustainable Development Strategies of Agro-Pastoral Transition Zone. 1st editor. Beijing, China: Chemical Industry Press; 2004. p. 102. ISBN: 7502554998

Section 2

Water Dynamics and Economics

Chapter 7

Economic Instruments to Combat Eutrophication: A Survey

Jean-Philippe Terreaux and Jean-Marie Lescot

Additional information is available at the end of the chapter

http://dx.doi.org/10.5772/intechopen.79666

Abstract

Eutrophication of aquatic ecosystems is a functional process triggered by excessive nutrient inputs into water courses. It causes disruption to ecosystems, with impacts on associated goods and services, which consequently might not be provided in a sustainable way. These impacts have served to politicize the issue in recent years. In this chapter, we present the main lessons learned from an international literature review on the economic aspects of eutrophication, first with the purpose of managing the problem in France and second in the context of a European research project. This study aims to help public decision-making in the reduction of this water pollution. By analyzing past experiences and the results of recent modeling work, it allows to avoid a number of pitfalls and focus on efficient solutions.

Keywords: economics, eutrophication, regulation, incentive, public policy

1. Introduction

Natural environments are no longer able to assimilate without harming all the pollution caused by human activities. Many rivers, coasts, and water bodies suffer from eutrophication [1, 2]. While the induced costs are difficult to estimate, they must be taken into account in public policies relating to agricultural and urban development. Eutrophication is triggered by excessive nutrient inputs, mainly nitrogen and phosphorus [3], causing increased levels of biomass in aquatic ecosystems. This can result in major disruption to aquatic ecosystems and may also impact associated goods and services, economic activities, and human health. The main sources of this pollution are agricultural activities, discharge from urban waste water treatment plants, and individual sewage treatment systems. The principal economic issues are the following: What is the best way to define and implement acceptable trade-offs by different stakeholders? How can economic activities and eutrophication be balanced in urban and rural territories while respecting the principles of sustainable development in a context of global change?

France is one of the countries affected by this phenomenon [1]. In view of this, the French government asked various research centers to carry out a literature review on the nature of the eutrophication, its causes and consequences, and potential mitigation measures. A total of 4000 documents (books, peer-reviewed articles) were analyzed in early 2017. In this article, we present part of this work, focusing exclusively on the economic aspects of public policy relating to this problem. For this purpose, 932 articles were selected from the Econlit and Scopus databases. Only the 382 most relevant of those were selected, following a review of their abstracts. More recent works were added later in the context of the Collaborative Land-Sea Integration Platform (COASTAL) research project.

In this chapter, we will focus exclusively on methodological works, using examples from case studies to illustrate a number of points. This will allow us to learn valuable lessons concerning the possible tools that could be developed for public decision-makers. In Section 2, we present general issues surrounding eutrophication prevention, namely, difficulty in defining a clear objective, difficulty in carrying out cost-benefit analyses, and the associated uncertainty and irreversibilities. We will also examine the consequences of combined pollution, both in terms of causes and effects. In Section 3, we explore possible ways of reducing pollution, followed by a more detailed presentation of the tools that can be used to deal with both diffuse and point source pollution in agricultural and domestic areas. The conclusion, in Section 4, summarizes the main lessons that can be learned from this work.

2. Difficulties in combatting eutrophication

2.1. Defining objectives

First of all, objectives have to be defined: without this first stage, it is difficult to rank possible actions and to subsequently evaluate the efficiency of policies. As shown by Naevdal in [4], there is generally an optimal level of eutrophication, which is neither the search for a total absence of eutrophication, which would involve too great a cost for society (lack of economic activities, high purification of water), nor the acceptance, without seeking improvement, of harmful levels of eutrophication. From an economic point of view, the most effective control of a pollutant is achieved when the marginal abatement costs are equal among all those responsible for discharges and when these costs are equal to the marginal benefit of a better water quality (see also Iho et al. in [5]).

That said, in most cases, information on marginal benefits is not available, and biophysical sciences (e.g., natural sciences) will set emission reduction targets based on environmental motives. In this case, the problem is how to achieve (for the best price) a given level of total discharge (or a water quality level), which is agreed upon through political channels. Further complicating the picture, eutrophication is most often linked to threshold effects (concentration levels for different pollutants), and once these thresholds are exceeded, ecosystem dynamics evolve, making it difficult to define optimal policies.

In this situation, as Xepapadeas observes in [6], it is better to couple ecological models and economic models. Such a representation makes it possible to take into account and combine elements such as strategic interactions between economic agents, non-convexities induced by nonlinear loops, different spatial and temporal scales, and the representation of different spatial and temporal dynamics. However, this may require the implementation of complex models. A new form of arbitration must then be found between the simplicity of representations and their realism. Neglecting the phenomena of bifurcation or irreversibility can thus lead to economically or ecologically undesirable states.

2.2. Cost-benefit analyses

Two objectives can be pursued: maximizing the net benefit of actions or minimizing costs with a given objective (see, e.g., Gren in [7] for a comparison of two possibilities applied to the Baltic Sea). Bryhn et al. insist in [8] that the costs of actions foreseen must in any case be compared to the expected benefits: for example, the Baltic Sea Action Plan signed in 2007 appears to have costed € 3 billion per year. It is then important to minimize the risk of waste of such big sums for more or less effective measures.

However, Huppes stressed in [9] that the direct and/or indirect costs of environmental policies are quite complex to define and calculate. For the latter, for example, public authorities bear the costs of control, the disputes arising from them, the costs of research needed for effective actions, companies bearing the costs of constraints, administrative costs, litigation costs, etc.

Additionally, these costs generally have a dynamic aspect (variation over time) that further complicates decision-making. Finally, transaction costs (negotiation costs, consultation costs, system administration costs, decision-making costs, etc.) are often far from insignificant but depend on intervention by public authorities, especially for diffuse pollution (see, e.g., McCann and Easter in [10]).

For policies relating to the agricultural community, von Blottnitz et al. recall in [11] that the way in which policies are implemented also impacts agricultural employment and business linked to the sector, as well as income and production for farmers (see Arata et al. in [12] for an example of reduction of livestock).

It is also necessary to determine who will bear these direct and indirect costs: Modifying certain practices (conditions of the use of fertilizers in agriculture, crop rotation as studied by Power et al. in [13], wastewater treatment modalities) will be reflected in prices (of agricultural products) and taxes (for local water purification) and may also result in a modification of the risks incurred (e.g., risk on the level of the agricultural income, on the quantity of production of food products, with repercussions going beyond the prices). In cost-benefit analyzes, it is essential not to put more emphasis on the present costs but to take better account of the long-term benefits to the environment through a judicious choice of the discount rate that must also incorporate uncertainties (see Ludwig et al. in [14]).

As regards the benefit side, the task is not easy either. For example, the assessment of the environmental or social benefits linked to less eutrophication are the subject of numerous papers (see [1]); but due to a lack of space, we cannot develop this aspect here.

The challenge is to find a method to fight eutrophication that is either incentivizing or binding and which will result in an acceptable balance for all parties (farmers, taxpayers, those benefiting from water quality, etc.), while taking into account the constraints that apply to everybody (agricultural markets fluctuating, overall tax burden compared to other countries, etc.).

2.3. Uncertainties, irreversibility, and robustness of solutions

The results of the various studies are generally derived from models or reasoning subject to numerous uncertainties. Some uncertainties affecting decision-making are described by Singh et al. in [15]: it may take the form of the impossibility of defining a single probability distribution for the most important parameters for the underlying model or to have a single well-defined objective to capture the simultaneous and divergent interests of the main stakeholders. This was already highlighted by Wladis et al. in [16], although it is often ignored by decision-makers.

Turner et al. in [17] also emphasize this fact in a context of scientific uncertainty. One management objective may be to maintain a certain stability of the environment, with parameters remaining within certain limits.

Lempert and Collins in [18] work in a completely different context, which involves making a decision in uncertainty when the links between actions and their consequences are relatively unknown. No attempt is made to seek the optimality of the solution in the context of the assumptions made and of the supposed value of the parameters. The objective is to have a solution that may be less efficient but more robust, namely, less sensitive to assumptions and satisfactory for a relatively wide range of future parameters and conditions, while keeping some options open.

Another difficulty is how to take into account fluctuations in pollutant emissions over time and not just take into consideration average values. This can lead to the simultaneous introduction of a number of different instruments, each pursuing a certain goal (reduction of average pollution, peak pollution).

An adaptive management model is described by Bond and Loomis in [19], where agents use small-scale experiments to test assumptions about global system responses. It is therefore necessary to arbitrate between collecting information and managing the system to achieve the objective (e.g., to move toward an optimal level of pollution). Agents can thus voluntarily deviate from the optimum trajectory for this purpose. Generally, it is understood that this method leads to better and more informed decisions when there are significant uncertainties.

It was within this framework that Ludwig et al. in [20] implemented a profit optimization model related to agricultural activities minus the costs associated with the eutrophication of a downstream lake. They show that the interaction of slow and fast variables can create resilient or vulnerable systems. To manage such a system, the solution may be to monitor appropriate slow variables and take action before it is too late. An approach based on quasi-option values (see Henry in [21]) would lead to a reduction in pollutants so as to remain below this possible limit that would induce a changeover. This is the case here with the potentially slow dynamics of phosphorus in sediments.

2.4. Cross aspects of pollutions

There are many sources of eutrophication, and these sources may interact or have cross impacts. Gren et al. show in [22] how simultaneously taking into account nitrogen and phosphorus to control pollution in the Baltic Sea reduces the overall cost of abatement by about 15% compared to a separate approach.

Kuosmanen and Laukkanen recall in [23] that reducing pollution requires a compromise between the reductions of these different pollutants. For example, the Helsinki Convention set 50% reduction targets for nitrate and phosphate emissions to combat eutrophication in the Baltic Sea. From an economic point of view, there is no reason to expect such uniform rates to produce a socially optimal reduction.

Given a source of pollution, its effects can take different forms, and it is preferable to take all of them into account in calculations. For example, von Blottnitz et al. indicate in [11] that the effects of the use of nitrogen fertilizers are climate change due to the production of these fertilizers, other pollutants emitted into the atmosphere during this production, the greenhouse effect induced by the application of fertilizers, eutrophication, drinking water pollution, and damage due to the emission of volatile substances (especially NH3) from these fertilizers.

In addition, Brink et al. describe in [24] how emissions of one pollutant may have impacts (positive or negative) on emissions of other pollutants. Some of them have local effects (e.g., on the eutrophication of rivers), while others have only an overall effect (greenhouse gases). These indirect effects, as well as their local or global nature, are most often ignored by decision-makers, whereas taking them into account would reduce the total cost of environmental protection, for a given objective.

3. Means to reduce water pollution

3.1. Different tools

Different instruments can be used to reduce water pollution: generally they will consist in incentives, regulations, physical facilities (e.g., buffer zones), or a combination of these. Within the framework of a "command and control" system, the regulator indicates the technical measures that should be taken and verifies that they are effective. For example, different types of standards can be related to agricultural inputs or individual treatment systems. Several problems emerge, all more or less linked to the information that the regulator has on the effectiveness of the measures imposed, the reality of their implementation, the diversity of local situations, and their impact on that effectiveness.

Latacz-Lohman and Hodge showed in [25] how the first generation of European agri-environmental measures have used this method, for example, with dates and concentrations of livestock manure application on agricultural land, while more recently market instruments have been put in place. However, as early as 1998, Cowan insisted in [26] that economic instruments, such as those presented later in this chapter, generally have a better potential in terms of cost-effectiveness than command and control methods.

Setting taxes and subsidies is less prescriptive, since the economic agent can refuse the subsidy or agree to pay the tax and continue as before. It provides incentives for the implementation of environment-friendly measures or discourages certain actions. In the case of subsidies, or payments for environmental services such as those described by Ma et al. in [27], arises the question of the financing capacities, and of "who bears the costs in the end".

In addition, a general problem emerges: Is it legitimate to finance the reduction of pollution, and is it not contrary to the polluter-pays principle? One possibility is to require the polluter to satisfy certain constraints in order to receive subsidies for other objectives (e.g., different agricultural subsidies). This is not a question of funding pollution reduction, but of making it a prerequisite for public aid.

One of the problems generally observed is that these constraints, whose costs to public authorities seem to be low, are often not very targeted or too general in their definition, making them largely ineffective. On the other hand, a certain inequity is created between the beneficiaries of the subsidies, since the cost for satisfying the constraints is not always proportional to the amount perceived.

For sites of specific interest, for example, of great environmental value, it is possible not to pay the owner for the opportunities foregone in protecting the environment, but to make reprehensible the actions harmful for the environment. In other words, the right to property is now accompanied by a duty to protect the natural environment in which that property is located (see Latacz-Lohmann and Hodge in [25]), and payments are made only for positive actions in favor of the environment.

As Romstad points out in [28], these subsidies can also be used to set up buffer zones or to protect wetlands, thus benefitting biodiversity and landscapes and lending less weight to the impression that polluters are subsidized. As for the establishment of wetlands, Byström et al. show in [29], theoretically and with an application in southwestern Sweden, that because the source of pollution is random (seasonal and annual variations), the efficiency of the wetlands is then also random.

In a very different region such as the Mississippi Basin, Roley et al. study and compare in [30] the cost and effectiveness of different measures such as wetlands, intermediate crops, and ditches in reducing nitrogen leakage. It must be noted that the combination of these various means is quite possible. As regards the parameterization of such actions, the system may evolve gradually as direct and indirect effects are observed, as available techniques evolve, and as the level of general pollution develops.

These subsidies, along with taxes, can be applied to inputs such as fertilizers used, and in this case, they are often easier to set up (because of lower transaction costs). The main problem comes from the fact that what is harmful is the pollutant, whereas what is taxed or subsidized is an input, and between the two, there is a whole process of transformation, which can differ from one agent to another. In addition, the impacts of pollutants can be very different from one geographical location to another. It is quite understandable (see, e.g., Taylor et al. [31]) that there is no single optimum instrument for all farms and that the choice of an instrument remains largely dependent on resource conditions and production potentials that impact the costs of reducing pollution. Finally, in the agricultural sector, for example, inputs may differ from one farm to another: the use of chemical fertilizers will be taxed, while the use of fertilizers produced by the animals of the farm will not. Sometimes, it is useful to differentiate measures according to the activities or circumstances in which pollution is generated.

Tradable permits are another option, namely, permits to emit a pollutant that can be sold or bought on a market. These permits are either issued free of charge or initially sold by public authorities (entailing an additional cost for involved stakeholders). Von Blottnitz et al. outline in [11] the properties of this type of instrument. Gren and Elofssen present in [32] different variants, their potential interests, by applying these instruments to their case study (the Baltic Sea). This instrument is more flexible than a command and control system and does not require a lot of information about the polluter. With permits, pollution is immediately reduced, although the level needed for pollution reduction is initially unknown. But if the price of permits on the market is to be equal to the marginal cost of pollution reduction, the regulator should regularly adjust the quantity of permits issued until a socially acceptable situation is achieved. Similarly, Mitchell describes in [33] a system of permits for spreading poultry manure in the Illinois Basin.

It should be noted that spatial heterogeneity has important effects on the level of benefits that can result from the exchange of permits to pollute, and this dimension must be taken into account in implementing such a system (see Lankovski et al. in [34]).

In addition, Akao and Managi show in [35] the importance of taking into account inter-temporal aspects in order to have an efficient system. A free-rider phenomenon (i.e., some people benefit from the effort of others) can arise at different scales: at the macroeconomic level, for example, around the Baltic Sea, some countries may expect their neighbors to make the first efforts, thus diluting the overall impact of pollution. The same effect applies at the local level for activities or people within the same watershed.

3.2. Toolkits for diffuse pollution

3.2.1. General information

Nonpoint source pollution is defined by the fact that the emissions of each agent are not directly observable at a reasonable cost. Xepapadeas describes in [36] three possible methods for reducing domestic and agricultural diffuse pollution, which are difficult to regulate due to information asymmetry between the polluter (who understands the effort needed to reduce effluents and the associated costs) and the regulator (who does not know them) and the random aspects between the polluter's actions (e.g., manure spreading) and the pollution measured in downstream watercourses:

- The first is to consider that pollution is a function of certain production factors (inputs) and the developed instrument is a system of taxes, sometimes subsidies, to reduce these inputs. Rougoor recalls in [37] that the interest of a tax comes from the ease of implementing and the associated transaction costs that are generally low. Negative aspects come from the absence of targeting in the case of problems restricted to a local area, where the scope of application of the tax does not correspond to that of pollutant emitting and more generally to the risk of a competitiveness decrease of the agricultural sector since production costs increase.

- The second is to observe pollution, for example, downstream (at the outlet) of a small watershed, to set an acceptable threshold, and to implement an ambient tax, or a global fine paid by all potential polluters irrespective of actual pollution, when it cross the defined limit; a subsidy may also be awarded where the measure gives a result below the threshold. Compared to a more systematic method of taxation, the first aim is to have a more efficient action because it is adapted to a more accurate geographical area and, on the other hand, to introduce collective responsibility of the farmers or inhabitants concerned. Conditional voluntary contracts can thus be set up, giving that way interest to everyone to respect the contract (e.g., reduction of pollution against subsidies).
- The third is to establish, where it is feasible and cost-effective, a system to control individual pollution and to tax any inappropriate behavior or excessive pollution. This means transforming nonpoint source pollution to point source pollution, which is already the case, for example, for the control of individual septic tanks. It is also possible to allow the polluter to demonstrate the true level of effort he is willing to contribute by choosing from a set of possible contracts or subsidies, the most suitable one for him.

Each option has advantages and disadvantages: for example, measuring inputs can cause excessive information costs in addition to other costs to reduce pollution but is anyway fairer than ambient tax. The latter are an easy way for the regulator to move the problem to a lower geographical level, relying on social control that is more possible within a smaller group. This system of collective punishment remains however particularly unfair and thus may be unacceptable. For this reason, when the third possibility is reasonably possible, it is generally preferred. Otherwise, measurement of inputs, if not too expensive, is a good second choice.

3.2.2. Nonpoint source agricultural pollutions

Agriculture is often an important source of nonpoint source pollution that because of its characteristics should be tackled in a particular way. Generally, solving diffuse agricultural pollution problems cannot rely on any one single solution: Pretty recalls in [38] that agriculture is, by definition, multifunctional (in the sense that it produces different goods together) and possibly the source of different negative externalities but also positive ones (landscapes, carbon sequestration, limitation of floods, etc.). The variety of situations and problems to be solved leads to various deftly articulated solutions to encourage certain practices and to dissuade others, ranging from advice to regulatory or legal measures, and to the use of various economic instruments. As Saysel shows in [39], it is sometimes simply a matter of giving regular and relevant information to agents, for example, on the judicious use of fertilizers depending on the situation. For farmers, financial variables (notably income) are the main basis on which measures are adopted or refused. On the other hand, those located in the most at-risk areas for eutrophication are not necessarily the most likely to adopt them. Grammatikopoulou et al. in [40] have shown for Finland that it is more efficient to implement targeted measure, e.g., for farmers who are the most able to reduce their emissions, although the cost of implementation is therefore much higher. Fezzi et al. worked in [41] on the costs for farmers of different measures to reduce eutrophication: They show, on a case study of watershed in England, the impact of the choice of a measure, especially the variability of this impact from one farm to another. In parallel, they recall the importance of the heterogeneity of soils and agricultural practices on the effects of such measures, as did Konrad et al. in [42] for the Odense Fjord in Denmark.

In the case of the Baltic Sea, Turner et al. [43] show that if the reduction of inputs is the most effective measure, uniformity of reduction measures is not optimal because of different situations between basins.

It is therefore important to think carefully as much about the local application modalities as the choice of an instrument. Konrad et al. use in [43] a spatialized agro-economic model to estimate the effect of different measures (Fjord Odense watershed) and show in particular that geographically targeted measures can lead to high transaction costs, in comparison with uniform measures, if only to define and then monitor their implementation. Xu et al. show in [44] how a well-chosen land use change, and modified agricultural management strategies, may lead to an efficient phosphorus emission reduction.

Three types of incentives for preserving permanent grasslands or converting cultivated land into wetlands on the west coast of Sweden are compared by Gren in [45]: a lump sum payment for the areas concerned, a set of contracts from which the farmer will choose the most attractive for him (thus revealing information about his costs of preservation or conversion), or finally a mutual agreement negotiation that is generally eliminated because it is too time consuming. The choice between the two first possibilities depends on the form of the cost and benefit functions of the farm.

The use of nitrogen fertilizers is studied in [46] by Williamson in the United States. He reminds us that their use depends on the price of fertilizers, the cost of agricultural products, and the way in which farmers manage risks, along with their knowledge of the real need of crops for fertilizers. Hansen and Hansen develop in [47] an interesting method for controlling eutrophication induced by phosphorus pollution: rather than simply taxing phosphorus inputs, they suggest taxing the difference between imports of inputs and exports in the form of agricultural products. Although their model does not take into account hazards, it provides an interesting perspective, especially since it includes storage of phosphorus in soil.

A system of nonlinear taxation and subsidies for reducing agricultural nonpoint source pollution is described in [48] by Bontems et al. Farmers differ according to dimensions such as common knowledge (knowledge shared by a group of agents, in which everyone knows that they all share it), spreading areas, level of production, or private knowledge on the way to limit pollution for equal production. In this framework, the authors look for ways to compensate farmers who implement costly practices but lead to pollution reduction; a system of payments revealing private information on each farmer efficiency makes it possible to improve the effort distribution between farms to reduce pollution.

The payment of subsidies for the adoption of measures to reduce pollution may also be conditional on the outcomes. In that respect, Talberth shows in [49] that payment for grass strips under condition of performance is superior, in the sense that it allows to obtain a better reduction of nutrients for the same budget allocation.

It is also important to avoid falling victim to deadweight effects, with high-cost measures and questionable effectiveness. This is mentioned by Dupraz et al. [50], who note that there is often an advantage to be gleaned from putting in place measures aimed at avoiding limit effects, e.g., applying to a minimum proportion of the farm's surface area or a minimum of intensity. Their model shows that this risk is increased by the information asymmetry that exists between regulators and farmers.

Changes in plowing practices are much more difficult to encourage, as shown in [51] by Orderud and Vogt in a study on an area southeast of Oslo, Norway. For the authors, the solution is to increase farmers' knowledge on the environmental issues and on the phosphorus cycle so that farmers could understand the complexity of the process and not be discouraged by immediate inconclusive results.

Kling presents in [52] an agro-economic model linking land use and resulting nitrate and phosphate pollution. The model is applied on two watersheds feeding the Mississippi River (USA) and makes it possible to test the effects of intermediate crops aiming at reducing eutrophication. He stresses the advantage of linking this type of model with a representation of farmers' behavior.

Establishing drinking water catchment areas with farming constraints that are compulsory and not financially compensated, and other areas where proposed measures are voluntary and financially compensated, is an option examined in [53] by Osborn and Cook, with a case study in the island of Thanet in the Northeastern part of Kent (UK). The authors address the issue of scale when defining zones: it should be not too coarse, so as not to unnecessarily penalize agriculture, and not too refined, for example, only around drinking water catchments for effectiveness purposes. Similarly, Balana et al. in [54], using an environmental, agronomic, and economic model applied to an agricultural area in eastern Scotland, determine the costs and effectiveness of implementing buffer zones along watercourses. They show, on the one hand, that for the same effectiveness in terms of phosphorus reduction, induced costs can be reduced by about 20% just by varying the width of these zones, rather than imposing buffer zones of a uniform width. They show that costs increase exponentially as a function of the amount of nitrate withdrawn.

3.2.3. Domestic diffuse pollution

Withers et al., working on five case studies in Europe (England, Ireland (two cases), Scotland, Norway), show in [55] that the number of individual treatment systems for domestic wastewater such as septic tanks is generally undervalued, which in turn makes other potential sources of eutrophication, in particular agriculture, responsible for the pollution observed.

Beyond the number, the performance of this type of treatment system is difficult to assess, because of a lack of information on their technical characteristics (implantation, age, level of maintenance, proximity to a watercourse, etc.). Although these systems often represent a small part of the nutrient load (mostly less than 10% on annual average in case studies), they can provide significant concentrations during certain periods, particularly in summer, and periods of low water. Increased owners' awareness of the need to properly maintain their sewage treatment facilities can be a fairly effective and inexpensive way to substantially improve the situation.

Motivations of Swedish owners for changing their individual treatment facilities, for achieving a better sewage treatment in order to reduce eutrophication, are examined in [56] by Wallin et al. Owners are motivated more by broader benefits (e.g., an improved functioning of their treatment system) and by fairness relative vis-à-vis other owners (that should not be exempted from the same changes), than by environmental concerns. In this context, economic incentives should work, while increased inspections would contribute to a sense of equity in addition to communication means on the merits of such changes.

3.3. Tools for point source pollution

Domestic, agricultural, or industrial sewage treatment plants are the main sources of point source water pollution, defined by the fact that the emissions of pollutants into the environment come from an identifiable source. Therefore regulations and incentives are generally easier to implement than in the case of diffuse pollution.

Moreover, wastewater can be reused for agriculture or green areas, in compliance with current water quality regulations. In this context, Verlicchi et al. provide in [57] a cost-benefit analysis of the implementation of a post-treatment zone of sewage by planted filters for the city of Ferrara (Italy). Overall, the result of the cost-benefit analysis is positive, despite the use of a discount rate of 5%. It is particularly interesting from an environmental point of view, by reducing the effluent discharged into the watercourse, and also at the urban level by the creation of recreational spaces.

With respect to wastewater treatment, Piao et al. compare in [58] different ways of treating sludge from sewage treatment plants and their indirect effects on eutrophication but also on their potential for global warming, toxicity to humans, and acidification of the natural environment. The incineration of sludge, with ultimate waste treatment, seems to be the best method for these four dimensions.

3.4. Practical conditions for implementation of measures

In most cases, implementing measures is not straightforward because of the different stakeholders involved and the various sources of pollution. Löwgren describes in [59] a process of stakeholder consultation, as called for by the European Union, in the form of two meetings of 1 day, each with a few dozen people, on measures to be taken to combat eutrophication in a watershed in Sweden. The author indicates that the results obtained are not representative in any statistical sense. Farmers are commonly referred to as the guilty party in cases of eutrophication, and they tend to defend themselves by drawing attention to shared responsibility with other professionals. While it is relatively easy to identify the impacts of agriculture, it is much more difficult to assess their benefits: food production, support for certain types of biodiversity, cultural heritage, and open spaces are often seen as given, and farmers no longer draw credit from these externalities either in monetary terms or even in the context of this kind of reflection.

However while cooperation between residents, businesses, and farmers is particularly important in order to combat eutrophication, Iwasa et al. show in [60] from a general model that this can lead to complicated dynamics in the natural environment. This is because the willingness of each stakeholder to cooperate depends on the cooperation of others, as well as on the overall environmental concerns of society. In the model, two factors will affect the decision of each agent: the cost of action for the environment and social pressure. For a lake, social pressure will generally increase with the level of pollution. In total, there are different positive or negative return forces, involving potentially varied dynamics.

Should then decisions be decentralized? Elofsson, working on pollution in the Baltic Sea, recalls in [61] that regional agencies managing water in a basin generally have a more detailed knowledge of local conditions than structures operating at a wider geographical scale, such as the European Union for this case study. It would therefore be interesting to decentralize decisions when assessing the means of reducing eutrophication. On the other hand, there is a risk that regional agencies will act according to their own interests rather than those of the higher-level structure. The model developed actually shows, for the case of the Baltic Sea study, that this effect is not particularly marked.

On this basis, what is the best way to decentralize? Kroiss examines in [62] various empirical strategies for protecting the Danube Basin and shows that much of the technical problems of water protection can be solved through national, regional, or local initiatives.

He mainly distinguishes two possible approaches: definition of an environmental standard or a precautionary method. The environmental standard indicates a minimum level of water quality, which must be satisfied everywhere, and the effectiveness of each treatment is thereby deduced, in particular from the dilution capacities of the natural environment. For the precautionary method, a minimum reduction in pollution or effluent quality must be achieved, regardless of the quality of the watercourse or its dilution or retention capacity.

Both approaches have their advantages and disadvantages (the precautionary approach being preferable for international problems, since the definition of environmental standards is more suited to the management of national basins or where the administration of a basin is centralized). In practice a combination of both would seem preferable. In fact, for the Danube, the main problem is not the translation of these methods in the texts but their actual implementation in practice.

4. Conclusion: lessons to be learnt from past experiences

In defining policies to combat eutrophication, the objective should be defined by models that simultaneously combine biophysical and economic aspects, and not by setting objectives for the state of the system, and then trying to minimize the costs of actions to be carried out. Other approaches can be taken, such as attempting to improve the current situation (e.g., Lake Apopka in Florida described by Fonyo and Boggess in [63]). For the economy, benefits must be compared to costs, whatever the tools used for the implementation of public decisions: command and control, regulation from taxes and subsidies, taxes depending on the results obtained locally ("ambient taxes"), and emission permits distributed free of charge or sold in auction. Uniform measures, such as a percentage reduction of emissions, are generally inefficient, and free-rider behavior is to be expected among some stakeholders.

Five main factors of the problem are often underestimated, as shown above or in practical examples described in the literature but which we do not have the place to present here:

1. The temporal dimension, with irreversibility in particular that may arise when crossing certain limits (e.g., a concentration level of a pollutant). This phenomenon can be taken into account with an appropriate representation of the systems.
2. Pollution often has several causes, and the choice to fight against one or several of them simultaneously or alternatively is far from being neutral on economic results. This is true both for the choice to take action against nitrogen and/or phosphorus in agriculture and to act preferentially or simultaneously on the domestic or agricultural sector or even to arbitrate between diffuse pollution and point source pollution.
3. Pollutions are often multiple (eutrophication, greenhouse effect, etc.), and efficiencies can be gained by taking that into account. Conversely, by taking into consideration the multiple benefits of reducing pollution, it is possible to consider alternatives that would otherwise be unprofitable.
4. The random nature of emissions modifies their effects in terms of eutrophication and may lead to the simultaneous introduction of different instruments. Furthermore, in this case, it may be preferable to seek certain robustness for these solutions, rather than optimality in one direction or the other. There will always be uncertainties, particularly those related to an imperfect knowledge of biophysical phenomena. It is not possible to wait until everything is known before acting. Adaptive management (by updating objectives, tools, or parameters, through experiments) can be a solution in this context.
5. The heterogeneous nature of sources, of the agents concerned, etc. cannot be neglected.

On the whole, it is not conceivable to copy a solution that has proved its worth in one context to solve another problem. On the other hand, lessons can be learned from successes or failures in very different situations. Ecological engineering solutions, apart from the development of buffer zones and wetlands, can have quite risky indirect effects. Sometimes, the question should be asked as to the comparative advantages of modest measures across large geographical areas or more substantial ones in smaller areas.

Finally, it should be noted that throughout this bibliographical analysis, the absence of ideal solutions and the interest of targeted policies designed for particular situations were highlighted. It is often case-driven instruments that can help to solve problems if they have been first properly identified and analyzed and if the solutions under consideration have been assessed in their different implications.

Acknowledgements

We thank the European Union for its financial support to the COASTAL project through the Horizon 2020 research and innovation program (grant agreement n° 773782) and the members of the French scientific expertise on eutrophication for the enriching exchanges. We also thank Sybille de Mareschal (Irstea, Clermont-Ferrand, France) for her assistance in finding the bibliographic references.

Conflict of interest

We have no pecuniary or other personal interests, direct or indirect, to declare in relation with the subject of this work.

Author details

Jean-Philippe Terreaux* and Jean-Marie Lescot

*Address all correspondence to: jean-philippe.terreaux@irstea.fr

Irstea Bordeaux, ETBX Research Unit, Cestas, France

References

[1] Gascuel CF, Moatar G. Pinay, Menesguen A, Souchon Y, Le Moal M, Levain A, Etrillard C, Pannard A, Souchu P, L'eutrophisation: Manifestations, causes, conséquence et prédictibilité, Quae, France. 2018. 128. p. ISBN 978-2-7592-2756-3

[2] Swain BR. Environmental challenges in the Baltic region: A perspective from economics, Palgrave, McMillan, London. 2017. 239 p. ISBN 978-3-319-56006-9

[3] Lwin CM, Murakami M, Hashimoto S. The implications of allocation scenarios for global phosphorus flow from agriculture and wastewater. Resources, Conservation and Recycling. 2017;**122**:94-105

[4] Nævdal E. Optimal regulation of eutrophying lakes, fjords, and rivers in the presence of threshold effects. American Journal of Agricultural Economics. 2001;**83**:972-984

[5] Iho A, Ahlvik L, Ekholm P, Lehtoranta J, Kortelainen P. Optimal phosphorus abatement redefined: Insight from coupled element cycles. Ecological Economics. 2017;**137**:13-19

[6] Xepapadeas A. Modeling complex systems. Agricultural Economics. 2010;**41**:181-191

[7] Gren I-M. Value of land as a pollutant sink for international waters. Ecological Economics. 1999;**30**:419-431

[8] Bryhn AC, Sessa C, Håkanson L. Costs, ecosystem benefits and policy implications of remedial measures to combat coastal eutrophication - a framework for analyses and a practical example related to the gulf of Riga. In: Eutrophication: Ecological Effects, Sources, Prevention and Reversal. 2010. pp. 103-134

[9] Huppes G. New instruments for environmental policy: A perspective. International Journal of Social Economics. 1988;**15**:42-50

[10] McCann L, Easter KW. Transaction costs of policies to reduce agricultural phosphorous pollution in the Minnesota river. Land Economics. 1999;**75**(3):402-414

[11] Von Blottnitz H, Rabl A, Boiadjiev D, Taylor T, Arnold S. Damage costs of nitrogen fertilizer in Europe and their internalization. Journal of Environmental Planning and Management. 2006;**49**:413-433

[12] Arata L, Peerlings J, Sckokai P. Manure market as a solution for the nitrates directive in Italy. New Medit: Mediterranean Journal of Economics, Agriculture and Environment. 2013;**12**(2):22-33

[13] Power JF, Wiese R, Flowerday D. Managing farming systems for nitrate control: A research review from management systems evaluation areas. Journal of Environmental Quality. 2001;**30**:1866-1880

[14] Ludwig D, Brock WA, Carpenter SR, Uncertainty in discount models and environmental accounting. Ecology and Society. 2005;**10**(2):13. [online] URL: http://www.ecologyand-society.org/vol10/iss2/art13/

[15] Singh R, Reed PM, Keller K. Many-objective robust decision making for managing an ecosystem with a deeply uncertain threshold response. Ecology and Society. 2015;**20**(3):12

[16] Wladis D, Rosen L, Kros H. Risk-based decision analysis of atmospheric emission alternatives to reduce ground water degradation on the European scale. Ground Water. 1999; **37**:818-826

[17] Turner RK, Bateman IJ, Georgiou S, Jones A, Langford IH. An ecological economics approach to the management of a multi-purpose coastal wetland. In: Working Paper—Centre for Social and Economic Research on the Global Environment. 2001. pp. 1-36

[18] Lempert RJ, Collins MT. Managing the risk of uncertain threshold responses: Comparison of robust, optimum, and precautionary approaches. Risk Analysis. 2007;**27**:1009-1026

[19] Bond CA, Loomis JB. Using numerical dynamic programming to compare passive and active learning in the adaptive management of nutrients in shallow lakes. Canadian Journal of Agricultural Economics. 2009;**57**:555-573

[20] Ludwig D, Carpenter S, Brock W. Optimal phosphorus loading for a potentially eutrophic lake. Ecological Applications. 2003;**13**:1135-1152

[21] Henry C. Option values in the economics of irreplaceable assets. Review of Economic Studies. 1974;**41**:89-104

[22] Gren IM, Savcavchuk OP, Jansson T. Cost-effective spatial and dynamic management of a eutrophied Baltic Sea. Marine Resource Economics. 2013;**28**:263-284

[23] Kuosmanen T, Laukkanen M. (In)efficient environmental policy with interacting pollutants. Environmental and Resource Economics. 2011;**48**:629-649

[24] Brink C, van Ierland E, Hordijk L, Kroeze C. Cost-effective emission abatement in Europe considering interrelations in agriculture. The Scientific World Journal [electronic resource]. 2001;**1**(Suppl 2):814-821

[25] Latacz-Lohmann U, Hodge I. European Agri-environmental policy for the 21st century. Australian Journal of Agricultural and Resource Economics. 2003;**47**:123-139

[26] Cowan S. Water pollution and abstraction and economic instruments. Oxford Review of Economic Policy. 1998;**14**(4):40-49

[27] Ma S, Swinton SM, Lupi F, Jolejole-Foreman C. Farmers' willingness to participate in payment-for-environmental-services programmes. Journal of Agricultural Economics. 2012;**63**(3):604-626

[28] Romstad E. The economics of eutrophication. In: Eutrophication: Causes, Consequences and Control. 2014;**2**:45-53

[29] Bystrom O, Andersson H, Gren I-M. Economic criteria for using wetlands as nitrogen sinks under uncertainty. Ecological Economics. 2000;**35**:35-45

[30] Roley SS, Tank JL, Tyndall JC, Witter JD. How cost-effective are cover crops, wetlands, and two-stage ditches for nitrogen removal in the Mississippi river basin? Water Resources and Economics. 2016;**15**:43-56

[31] Taylor ML, Adams RM, Miller SF. Farm-level response to agricultural effluent control strategies: The case of the Willamette valley. Journal of Agricultural and Resource Economics. 1992;**17**(1):173-185

[32] Gren I-M, Elofsson K. Credit stacking in nutrient trading markets for the Baltic Sea. Marine Policy. 2017;**79**:1-7

[33] Mitchell DM. An examination of non-regulatory methods for controlling nonpoint source pollution [thesis]. Oklahoma State University; 2001

[34] Lankoski J, Lichtenberg E, Ollikainen M. Point/nonpoint effluent trading with spatial heterogeneity. American Journal of Agricultural Economics. 2008;**90**(4):1044-1058

[35] Akao K-I, Managi S. A tradable permit system in an intertemporal economy. Environmental and Resource Economics. 2013;**55**(3):309-336

[36] Xepapadeas A. The economics of non-point-source pollution. Annual Review of Resource Economics. 2011;**3**:355-373

[37] Rougoor CW. Experiences with fertilizer taxes in Europe. Journal of Environmental Planning and Management. 2001;**44**(6):877-887

[38] Pretty J. Policy challenges and priorities for internalizing the externalities of modern agriculture. Journal of Environmental Planning and Management. 2001;**44**:263-283

[39] Saysel AK. Role of information feedback in soil nitrogen management: Results from a dynamic simulation game. Systems Research and Behavioral Science. 2017;**34**(4):424-439

[40] Grammatikopoulou I, Pouta E, Myyrä S. Exploring the determinants for adopting water conservation measures. What is the tendency of landowners when the resource is already at risk? Journal of Environmental Planning and Management. 2016;**59**(6):993-1014

[41] Fezzi C, Hutchins M, Rigby D, Bateman IJ, Posen P, Hadley D. Integrated assessment of water framework directive nitrate reduction measures. Agricultural Economics. 2010; **41**:123-134

[42] Konrad MT, Andersen HE, Thodsen H, Termansen M, Hasler B. Cost-efficient reductions in nutrient loads; identifying optimal spatially specific policy measures. Water Resources and Economics. 2014;**7**:39-54

[43] Turner RK, Georgiou S, Gren IM, Wulff F, Barrett S, Söderqvist T, Bateman IJ, Folke C, Langaas S, Zylicz T, Mäler KG, Markowska A. Managing nutrient fluxes and pollution in the Baltic: An interdisciplinary simulation study. Ecological Economics. 1999; **30**(2):333-352

[44] Xu H, Braun DG, Moore MR, Currie WS. Optimizing spatial land management to balance water quality and economic returns in a Lake Erie watershed. Ecological Economics. 2018;**145**:104-114

[45] Gren I-M. Uniform or discriminating payments for environmental production on arable land under asymmetric information. European Review of Agricultural Economics. 2004;**31**:61-76

[46] Williamson JM. The role of information and prices in the nitrogen fertilizer management decision: New evidence from the agricultural resource management survey. Journal of Agricultural and Resource Economics. 2011;**36**:552-572

[47] Hansen LB, Hansen LG. Can non-point phosphorus emissions from agriculture be regulated efficiently using input-output taxes? Environmental and Resource Economics. 2014;**58**:109-125

[48] Bontems P, Rotillon G, Turpin N. Self-selecting Agri-environmental policies with an application to the Don watershed. Environmental and Resource Economics. 2005;**31**:275-301

[49] Talberth J et al. Pay for performance: Optimizing public investments in agricultural best management practices in the Chesapeake Bay watershed. Ecological Economics. 2015;**118**:252-261

[50] Dupraz P, Latouche K, Turpin N. Threshold effect and co-ordination of agri-environmental efforts. Journal of Environmental Planning and Management. 2009;**52**:613-630

[51] Orderud GI, Vogt RD. Trans-disciplinarity required in understanding, predicting and dealing with water eutrophication. International Journal of Sustainable Development and World Ecology. 2013;**20**:404-415

[52] Kling CL. Luminate: Linking agricultural land use, local water quality and Gulf of Mexico hypoxia. European Review of Agricultural Economics. 2014;**41**:431-459

[53] Osborn S, Cook HF. Nitrate vulnerable zones and nitrate sensitive areas: A policy and technical analysis of groundwater source protection in England and Wales. Journal of Environmental Planning and Management. 1997;**40**:217-233

[54] Balana BB, Lago M, Baggaley N, Castellazzi M, Sample J, Stutter M, Slee B, Vinten A. Integrating economic and biophysical data in assessing cost-effectiveness of buffer strip placement. Journal of Environmental Quality. 2012;**41**:380-388

[55] Withers PJA, May L, Jarvie HP, Jordan P, Doody D, Foy RH, Bechmann M, Cooksley S, Dils R, Deal N. Nutrient emissions to water from septic tank systems in rural catchments: Uncertainties and implications for policy. Environmental Science and Policy. 2012; **24**:71-82

[56] Wallin A, Zannakis M, Johansson LO, Molander S. Influence of interventions and internal motivation on Swedish homeowners' change of on-site sewage systems. Resources, Conservation and Recycling. 2013;**76**:27-40

[57] Verlicchi P, Al Aukidy M, Galletti A, Zambello E, Zanni G, Masotti L. A project of reuse of reclaimed wastewater in the Po Valley, Italy: Polishing sequence and cost benefit analysis. Journal of Hydrology. 2012;**432**(433):127-136

[58] Piao W, Kim Y, Kim H, Kim M, Kim C. Life cycle assessment and economic efficiency analysis of integrated management of wastewater treatment plants. Journal of Cleaner Production. 2016;**113**:325-337

[59] Löwgren M. The water framework directive: Stakeholder preferences and catchment management strategies—Are they reconcilable? Ambio: A Journal of the Human Environment. 2005;**34**:501-506

[60] Iwasa Y, Uchida T, Yokomizo H. Nonlinear behavior of the socio-economic dynamics for lake eutrophication control. Ecological Economics. 2007;**63**:219-229

[61] Elofsson K. Climate change and regulation of nitrogen loads under moral hazard. European Review of Agricultural Economics. 2014;**41**:327-351

[62] Kroiss H. Water protection strategies-critical discussion in regard to the Danube river basin. Water Science and Technology. 1999;**39**(8):185-192

[63] Fonyo CM, Boggess WG. Coordination of public and private action. A case study of lake restoration. Water Resources Bulletin. 1989;**25**:309-317

Chapter 8

Setting Up a Computer Simulation Model in an Arkansas Watershed for the MRBI Program

Gurdeep Singh and Mansoor Leh

Additional information is available at the end of the chapter

http://dx.doi.org/10.5772/intechopen.80902

Abstract

The Mississippi River Basin Healthy Watersheds Initiative (MRBI) program launched by the USDA Natural Resources Conservation Service (NRCS) aims to improve the water quality within the Mississippi River Basin. Lake Conway Point Remove (LCPR) watershed, being one of the MRBI watersheds, is a potential candidate for evaluating the effectiveness of MRBI program. Recommended best management practices (BMPs) for LCPR watershed are pond, wetland, pond and wetland, cover crops, vegetative filter strips, grassed waterways, and forage and biomass planting. Before simulating these practices, it is essential to prepare the data needed for model setup to avoid the issue of garbage in, garbage out. This chapter focuses on detailed steps of preparing the data for model setup along with the calibration and validation of the model. The calibration and validation results were within the acceptable bounds. The results from this study provide the data to help simulate the MRBI best management practices effectively and prioritize monitoring needs for collecting watershed response data in LCPR.

Keywords: best management practices, modeling, water quality, SWAT, MRBI

1. Introduction

The Mississippi River Basin Healthy Watersheds Initiative (MRBI) program aims at implementing best management practices (BMPs) to control water quality. Quantifying the impacts of BMPs is important to demonstrate the worth of the MRBI program. Out of various MRBI-selected watersheds, the Lake Conway Point Remove (LCPR) watershed is the one listed in the 2011–2016 priority watershed by the Arkansas Natural Resources Commission (ANRC) [1, 2].

Field studies can be laborious and time-consuming; therefore, watershed modeling technique is generally used for analyzing the effects of BMPs on water quality. The Soil and Water Assessment Tool (SWAT, [3]) model was selected for this study. The SWAT model has been widely applied across the globe to assess the impact of various BMPs [4]. SWAT has also been applied to various watersheds in Arkansas—L'Anguille River Watershed [5, 6], Cache River Watershed [7], and Illinois River Watershed [8]. SWAT allows modifications of various parameters to simulate BMPs [9] and was applied at various spatial and temporal scales [10]. SWAT has been used to simulate impacts of land uses and BMPs [11, 12], develop maximum daily load plans [13, 14], and evaluate impacts on water quality [15, 16]. However, before simulating BMPs, it is essential to acquire and process the data needed for setting up a good model.

The goal of this chapter is to describe the steps in detail for acquiring and processing the data needed to set up, calibrate, and validate the SWAT model for the LCPR watershed.

2. Methodology

2.1. Study area

The Lake Conway Point Remove (LCPR) watershed is a 2950 km^2 (1140 miles^2) watershed located in central Arkansas within the counties of Conway, Faulkner, Perry, Pope, Pulaski, Van Buren, and Yell (**Figure 1**). The watershed has mixed land uses of forest, pasture, urban, and

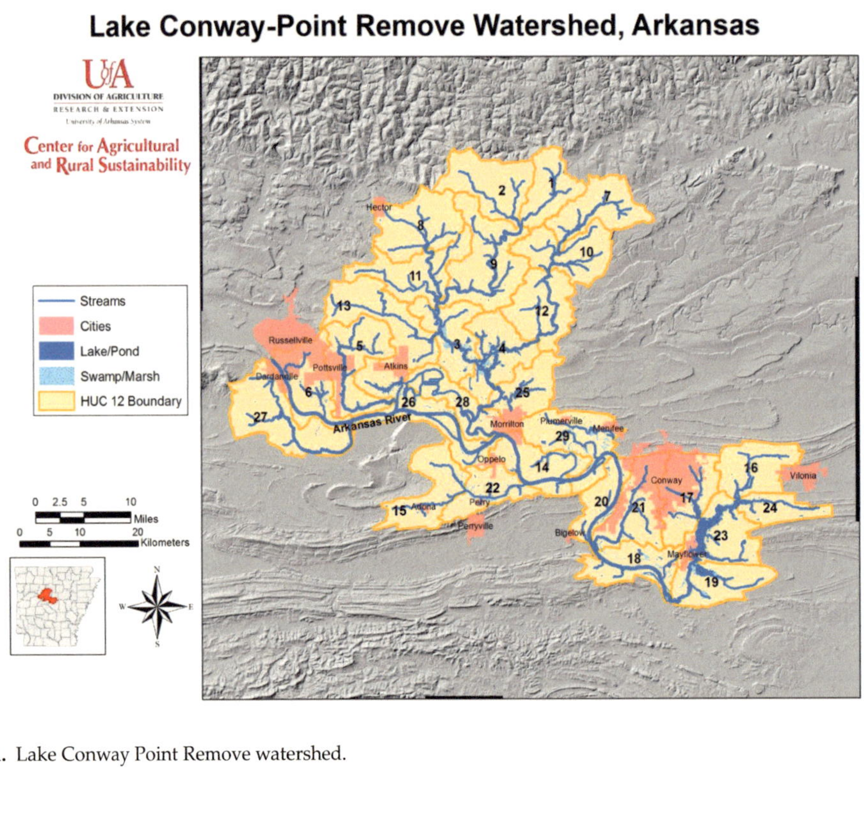

Figure 1. Lake Conway Point Remove watershed.

http://dx.doi.org/10.5772/intechopen.80902

cropland. An increase in urbanization, in parts of the watershed, has occurred since 1999. The subwatersheds within LCPR along with the area and hydrological unit codes (HUC) can be seen in **Table 1**.

2.2. Data preparation

The objective of this task was to collect and organize all data needed for the SWAT model setup at a 12-digit hydrological unit code within the LCPR watershed. Geospatial, watershed management,

Subwatershed	Subwatershed name	Area (km^2)	HUC no.
1	Trimble creek-west fork point remove creek	77.0	111102030102
2	Brock creek	113.1	111102030101
3	Devils creek-west fork point remove creek	88.2	111102030107
4	Barns branch-east fork point remove creek	102.7	111102030204
5	Galla creek	118.0	111102030303
6	Whig creek-Arkansas river	106.3	111102030302
7	Mountain view-east fork point remove creek	97.8	111102030201
8	Upper clear creek	120.4	111102030103
9	Rock creek-west fork point remove creek	156.2	111102030105
10	Sunny side creek-east fork point remove creek	100.9	111102030202
11	Lower clear creek	106.5	111102030104
12	Prairie creek-east fork point remove creek	106.9	111102030203
13	Gum log creek	130.4	111102030106
14	Portland bottoms-Arkansas river	90.9	111102030503
15	Headwaters rocky Cypress creek	100.1	111102030501
16	Jim creek-Palarm creek	92.4	111102030402
17	Little creek-Palarm creek	106.8	111102030403
18	Beaverdam creek-Arkansas river	88.0	111102030507
19	Little Palarm creek-Palarm creek	89.9	111102030405
20	Taylor creek-Arkansas river	65.1	111102030506
21	Tupelo bayou	110.8	111102030505
22	Outlet rocky cypress creek	70.5	111102030502
23	Pierce creek-Palarm creek	100.0	111102030404
24	Little cypress creek-Palarm creek	53.4	111102030401
25	Overcup creek	81.1	111102030205
26	Khun Bayou-Arkansas River	131.1	111102030304
27	Long Lake-Harris creek	148.2	111102030301
28	Point remove creek	80.2	111102030206
29	Miller Bayou-Arkansas river	116.4	111102030504

Table 1. List of HUC 12 subwatersheds and area in LCPR watershed.

water quantity, and point source data that were available and usable at the time of modeling were collected and reorganized in a consistent format for use in the SWAT model.

2.2.1. Elevation

The elevation dataset was retrieved at a 5 m resolution from GeoStor. This 5 m dataset was resampled to a 10 m resolution to reduce the size of huge files and increase the computation efficiency. The elevation map for LCPR can be seen in **Figure 2**.

2.2.2. Soils

The soil data were acquired from the Soil Survey Geographic (SSURGO) database for all LCPR counties in Arkansas and combined to make a soil map for the entire watershed. The SSURGO is the most comprehensive and detailed soil dataset available for LCPR. The soil map for LCPR can be seen in **Figure 3**.

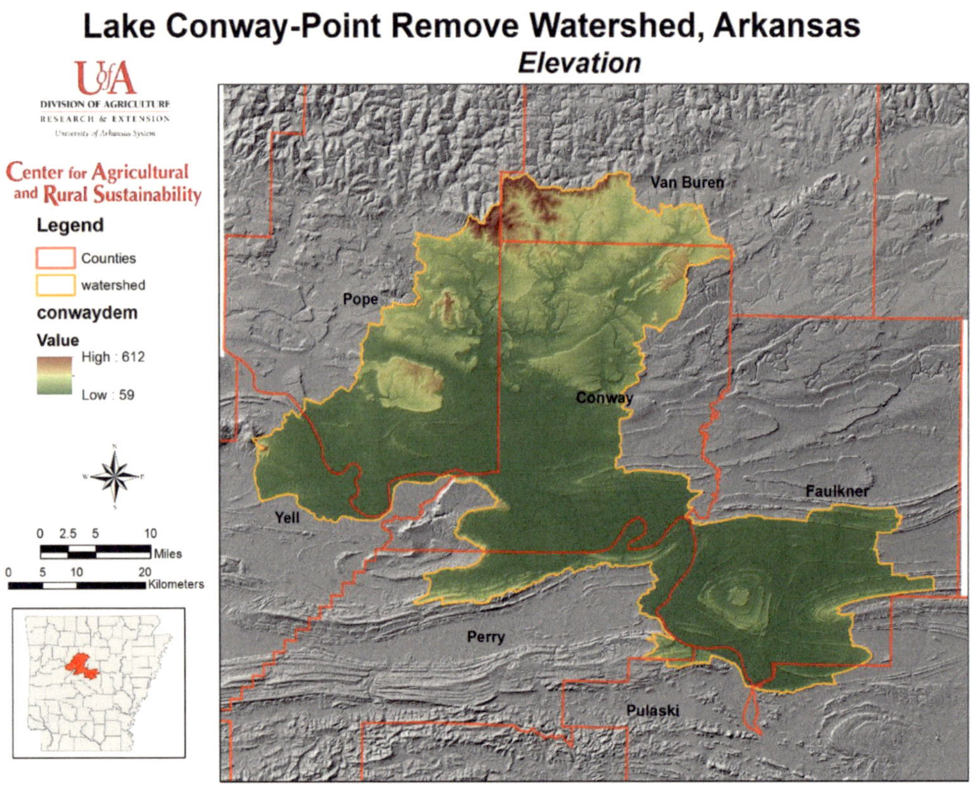

Figure 2. Lake Conway Point Remove watershed elevation.

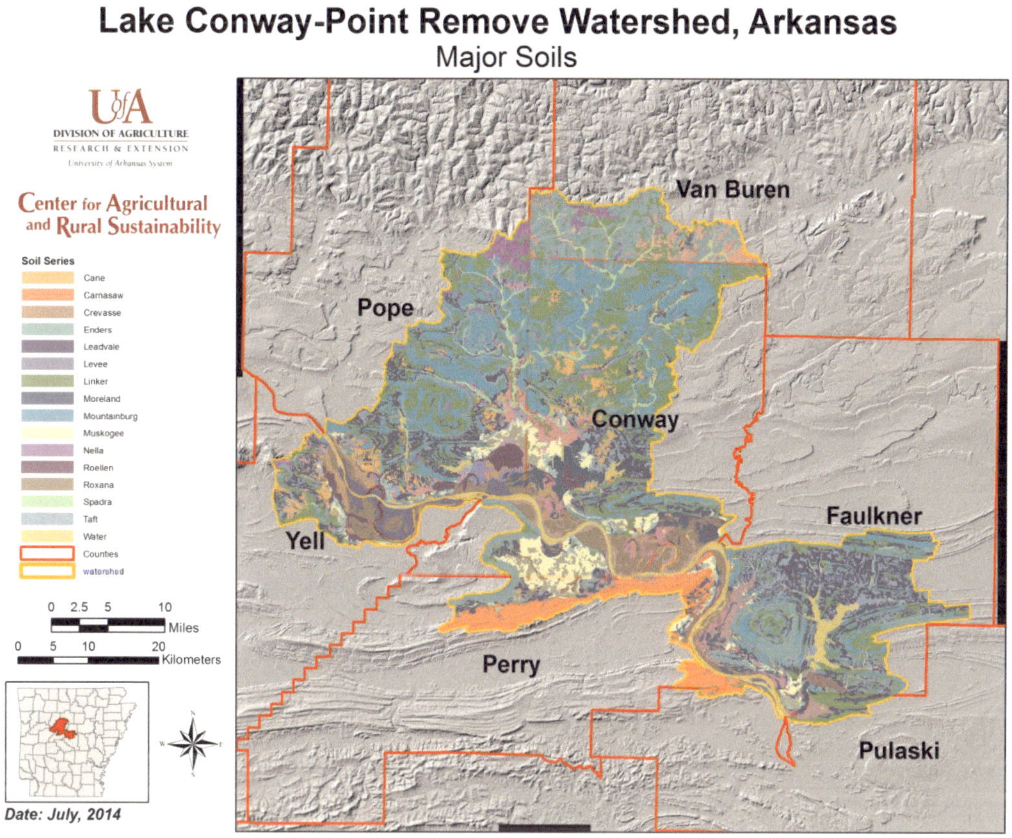

Figure 3. Soil map of Lake Conway Point Remove watershed, Arkansas, showing major soil series.

2.2.3. Land use/land cover

Land use and land cover data were acquired for 1999, 2004, and 2006 from GeoStor. Forest area was observed to be the most dominant land use and cover in the LCPR watershed. All land use and land covers were reclassified to make it compatible with the SWAT model. The land use and land cover map for LCPR can be seen in **Figure 4**.

2.2.4. Climate

Climatic data specifically daily precipitation and maximum and minimum temperature data were obtained from 90 climate stations from the NOAA's National Climatic Data Center (NCDC). Data are available from 1980 to 2012 for at least one of the climatic parameters. The procedure recommended by USDA-ARS in developing SWAT-formatted climate data were followed. Daily climate data were obtained using an inverse distance-weighted interpolation algorithm. The average data were calculated for each subwatershed using a pseudo-weather

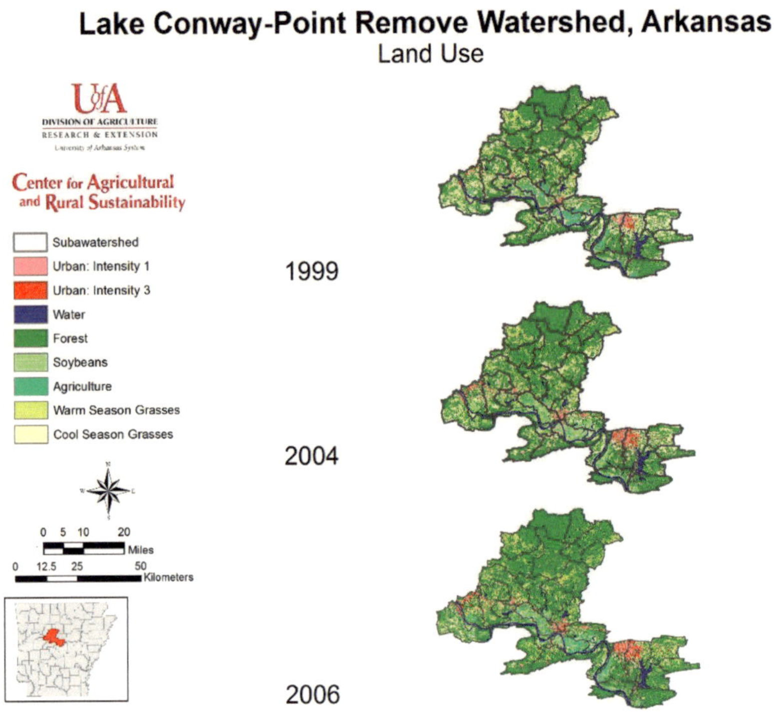

Figure 4. Land use and land cover in the Lake Conway Point Remove watershed.

station. NCDC validation results at each calibration station using leave-one-out cross-validation technique can be seen in **Table 2**. NEXRAD data were obtained from the Arkansas Basin River Forecasting Center (ABRFC).

2.2.5. Streamflow

The flow data are available for the West Fork Point Remove Creek near the Hattieville monitoring station from the US Geological Survey (USGS). This monitoring station is located in subwatershed 3 and covers approximately 20% of LCPR. The flow data were split between surface and baseflow using the baseflow filter program by [17].

2.2.6. Point sources

Point source data were obtained from the Arkansas Department of Environmental Quality (ADEQ) and was processed in the SWAT-compatible format. Point source data were available for flow, total suspended solids, organic nitrogen, organic and mineral phosphorus, nitrate nitrogen, ammonia nitrogen, and carbonaceous biochemical oxygen demand (CBOD). Locations for active point source facility that was incorporated in the SWAT model can be seen in **Table 3**.

Station	Parameter	DRAIN[1]	DNO_RAIN[2]	ME[3]	d[4]	PBIAS[5]%	R2[6]	NSE[7]	MAE[8]	RMSE[9]
Center Ridge, 4.S, AR, USA	PRCP	0.94	0.86	−0.12	0.95	−0.3	0.83	0.83	15.48	45.03
Conway, AR, USA	PRCP	0.91	0.79	−0.64	0.87	−1.9	0.59	0.58	23.53	63.56
Dardanelle, AR, USA	PRCP	0.95	0.79	0.51	0.85	1.5	0.54	0.52	24.55	71.4
Hattieville, AR, USA	PRCP	0.95	0.82	0.08	0.92	0.2	0.74	0.73	18.13	57.15
Morrilton, AR, USA	PRCP	0.90	0.82	0.97	0.9	2.8	0.69	0.68	19.84	59.78
North Little Rock Airport, AR, USA	PRCP	0.90	0.81	0.23	0.85	0.7	0.56	0.55	24.37	69.37
Perry, AR, USA	PRCP	0.90	0.82	−1.19	0.89	−3.3	0.65	0.64	21.71	64.82
Russellville Municipal Airport, AR, USA	PRCP	0.68	0.84	1.85	0.67	5.9	0.24	0.03	34.7	99.07
Conway, AR, USA	TMAX			0.45	0.99	0.2	0.95	0.95	14.49	22.31
Dardanelle, AR, USA	TMAX			−5.02	0.99	−2.2	0.95	0.94	15.14	22.95
Morrilton, AR, USA	TMAX			−1.9	0.99	−0.8	0.94	0.94	17.39	23.86
North Little Rock Airport, AR, USA	TMAX			4.05	1	1.8	0.99	0.99	9.03	11.83
Russellville Municipal Airport, AR, USA	TMAX			2.42	0.99	1	0.95	0.95	13.71	22.57
Conway, AR, USA	TMIN			−7.55	0.98	−7.1	0.95	0.94	15.59	22.75
Dardanelle, AR, USA	TMIN			−7.89	0.99	−7.8	0.95	0.95	14.18	21.36
Morrilton, AR, USA	TMIN			5.27	0.98	5.7	0.94	0.94	15.89	23.35
North Little Rock Airport, AR, USA	TMIN			−9.94	0.99	−8.3	0.97	0.95	14.79	19.68
Russellville Municipal Airport, AR, USA	TMIN			6.76	0.99	6.9	0.96	0.95	13.11	20.5

[1]NEXRAD detection conditioned on exceeding a given threshold gauge observations (DRAIN).
[2]NEXRAD detects no rainfall event (DNO_RAIN).
[3]Mean error (ME).
[4]Index of agreement (d).
[5]Percent bias (PBIAS).
[6]Coefficient of determination (R2).
[7]Nash-Sutcliffe efficiency (NSE).
[8]Mean absolute error (MAE).
[9]Root-mean-square error (RMSE).

Table 2. NCDC precipitation and minimum and maximum temperature validation results at each calibration station using leave-one-out cross-validation.

2.2.7. Cattle grazing, manure deposition, and poultry litter application

The detailed method for estimating pastures that should be receiving litter applications can be seen below.

No.	Subbasin	Facility	NPDES_ID	Latitude	Longitude
1	5	City of Pottsville	AR0048011	35.23	−93.05
2	6	City of Dardanelle	AR0033421	35.19	−93.14
3	6	Dardanelle water treatment plant	ARG640149	35.21	−93.15
4	6	Tyson Foods Inc., Dardanelle	AR0036714	35.22	−93.16
5	6	Russellville Water and Sewer System, City Corporation	AR0021768	35.25	−93.12
6	6	Freeman Brothers, Inc., d/b/a Bibler Brothers Lumber Company	AR0044474	35.25	−93.13
7	7	SEECO, Inc., J and R Farms SE1	AR0052221	35.43	92.56
8	7	Hamilton Aggregates	ARG500026	35.44	−92.54
9	8	Dover Water Works	ARG640148	35.40	−93.12
10	9	Quality Rock/Jerusalem Quarry	ARG500039	35.39	−92.80
11	10	KT Rock LLC	ARG500031	35.41	−92.67
12	11	SEECO, Inc., Campbell Thomas SE1	AR0052141	35.40	−92.83
13	13	City of Atkins	AR0034665	35.25	−92.92
14	14	Environmental Solutions and Services, Inc.	AR0051357	35.09	−92.71
15	14	Green Bay Packaging, Inc., Arkansas Kraft Division	AR0001830	35.10	−92.74
16	16	Rogers Group, Inc., Beryl Quarry	AR0047520	35.07	−92.25
17	16	Roy Nunn	ARG550322	35.07	−92.37
18	16	Waste Water Management, Inc. d/b/a Oak Tree Subdivision	AR0050792	35.08	−92.35
19	16	Fritts Construction, Inc., Hayden's Place Subdivision	AR0050253	35.09	−92.34
20	16	BHT Investment Company, Inc.	AR0044997	35.09	−92.33
21	16	Rolling Creek POA	AR0042536	35.11	−92.33
22	16	Genesis Water Treatment, Inc.	AR0051152	35.11	−92.34
23	17	Faulkner County Public Facility Board, d/b/a Preston Community WW Utility	AR0050571	35.03	−92.41
24	17	Wilhelmina Cove property owner	AR0048682	34.93	−91.11
25	17	City of Conway, Stone Dam Creek	AR0033359	35.05	−92.44
26	17	Coreslab Structures (ARK), Inc.	AR0050474	35.06	−92.43
27	17	MAPCO Express, Inc. #3059	AR0045071	35.07	−92.42
28	17	Flushing Meadows Water Treatment, Inc.	AR0048879	35.06	−92.37
29	17	Jesse Ferrel d/b/a Jesse Ferrel Rental Development	AR0049832	35.09	−92.37
30	18	City of Mayflower	AR0037206	34.95	−92.45
31	18	Carla Knight	ARG550430	34.97	−92.48
32	19	Construction Waste Management, Inc. Class IV Landfill	AR0051764	34.93	−92.44
33	19	Grassy Lake Apartments	AR0050334	34.94	−92.43
34	20	City of Bigelow	AR0049999	35.00	−92.61
35	20	City of Conway, Tucker Creek WWTP	AR0047279	35.07	−92.50

http://dx.doi.org/10.5772/intechopen.80902

No.	Subbasin	Facility	NPDES_ID	Latitude	Longitude
36	21	Conway Corporation, Tupelo Bayou WWTP	AR0051951	35.05	−92.54
37	22	City of Oppelo	AR0047643	35.08	−92.76
38	24	Faulkner County POID, Seven Point Lake Project	AR0050903	35.02	−92.18
39	25	Rogers Group, Inc.	ARG500066	35.24	−92.65
40	26	Lentz Sand and Gravel, LLC	ARG500072	35.12	−92.76
41	26	City of Atkins, South WWTP	AR0034673	35.22	−92.93
42	29	Rogers Group, Inc., Toad Suck Quarry	AR0047104	35.11	−92.56
43	29	City of Morrilton	ARG160001	35.13	−92.70
44	29	City of Menifee	AR0049361	35.14	−92.55
45	29	Gericorp, Inc.	AR0048623	35.15	−92.72

Table 3. Active point source facility location incorporated into the SWAT model.

Detailed methods for estimating pastures that received litter application:

1. Create buffer of a random radius around the active poultry houses.
2. Extract pasture areas under the buffer.
3. Assuming a grazing density of 1 cow/0.8 ha of litter amended pasture, calculate the number of cows that can fit the buffer.
4. Compare the calculated number of cows to the number of cows in the subwatershed.
5. Repeat steps 1–4 to obtain the best agreement between estimated numbers of cows.
6. Apply litter to pasture HRUs that fall under the best buffer radius.

The SWAT compatible data for cattle grazing, manure deposition, and poultry litter application can be seen in **Table 4**.

2.2.8. Urban pasture management

The pasture management schedule relating to specific operation and crop can be seen in **Table 5**.

2.2.9. Ponds and wetlands

SWAT input parameters relating to ponding were PND_FR, PND_PSA (ha), PND_PVOL (104 m^3), PND_ESA, PND_EVOL, and PND_VOL. These ponding parameters can be seen in **Table 6**. SWAT input parameters relating to wetland were WET_FR, WET_NSA (ha), WET_NVOL 104 (m^3), WET_MXSA (ha), WET_MXVOL 104 (m^3), and WET_VOL 104(m^3). These wetland parameters can be seen in **Table 7**.

Subbasin	Cattle grazing rate (kg/day/ha)	Cattle manure deposition rate (kg/day/ha)	Litter application/grazing
1	14.38	5.59	Yes
2	12.59	4.90	Yes
3	9.16	3.57	Yes
4	11.46	4.46	Yes
5	6.11	2.38	Yes
6	5.83	2.27	Yes
7	13.18	5.13	Yes
8	6.27	2.44	Yes
9	11.43	4.45	Yes
10	11.46	4.46	Yes
11	7.34	2.86	Yes
12	11.46	4.46	Yes
13	6.11	2.38	Yes
14	10.51	4.09	Yes
15	9.05	3.52	Yes
16	12.03	4.68	No
17	12.03	4.68	No
18	11.98	4.66	No
19	12.44	4.84	No
20	6.44	2.51	No
21	12.03	4.68	No
22	9.24	3.60	Yes
23	12.03	4.68	No
24	12.03	4.68	Yes
25	11.46	4.46	Yes
26	7.84	3.05	Yes
27	4.50	1.75	Yes
28	9.15	3.56	Yes
29	10.70	4.16	Yes

Table 4. Cattle grazing, manure deposition, and poultry litter application data incorporated into the SWAT model.

2.3. Model setup

SWAT divides a watershed into subwatersheds and further subwatersheds into hydrological response units. User-defined approach for delineating subwatersheds was used. ArcSWAT

Date	End	No. of days	Operation	Comment	Crop
				Cool-season grass (fescue)	
1-Apr			Fertilizer	Poultry litter@1 ton/acre of auto-fertilize	BERM
1-May			Planting	Warm-season grass (Bermuda)	BERM
15-May	31-Oct	170	Grazing		BERM
15-Jun			Hay cutting	85% removal	BERM
15-Jul			Hay cutting	85% removal	BERM
15-Aug			Hay cutting	85% removal	BERM
15-Sept			Hay cutting	85% removal	BERM
15-Oct			Hay cutting	85% removal	BERM
1-Mar			Fertilizer	Poultry litter@1 ton/acre of auto-fertilize	BERM
15-May	30-Oct	170	Grazing		BERM
15-Jun			Hay cutting	85% removal	BERM
15-Jul			Hay cutting	85% removal	BERM
15-Aug			Hay cutting	85% removal	BERM
15-Sept			Hay cutting	85% removal	BERM
15-Oct			Hay cutting	85% removal	BERM
1-Apr			Fertilizer	Poultry litter@1 ton/acre of auto-fertilize	BERM
				Warm-season grass (Bermuda)	
31-Aug			Fertilizer	Poultry litter@1 ton/acre of auto-fertilize	FESC
1-Sept			Planting	Cool-season grass (fescue)	FESC
15-Mar	1-Jun	79	Grazing		FESC
15-May			Hay cutting	85% removal	FESC
15-Jun			Hay cutting	85% removal	FESC
1-Sept			Fertilizer	Poultry litter@1 ton/acre of auto-fertilize	FESC
1-Oct			Grazing		FESC
15-Oct			Hay cutting	85% removal	FESC
21-Feb			Fertilizer	Poultry litter@1 ton/acre of auto-fertilize	FESC
15-Mar	1-Jun	79	Grazing		FESC
15-May			Hay cutting	85% removal	FESC
15-Jun			Hay cutting	85% removal	FESC
1-Sept			Fertilizer	Poultry litter@1 ton/acre of auto-fertilize	FESC
1-Oct	30-Nov	61	Grazing		FESC
21-Feb			Fertilizer	Poultry litter@1 ton/acre of auto-fertilize	FESC

Table 5. Pasture management schedule incorporated into the SWAT model.

Subwatershed	PND_FR	PND_PSA (ha)	PND_PVOL (104 m^3)	PND_ESA	PND_EVOL	PND_VOL
1	0.068	30	30	40	40	30
2	0.007	4	4	6	6	4
3	0.290	146	146	195	195	146
4	0.330	194	194	258	258	194
5	0.066	45	45	60	60	45
6	0.090	55	55	73	73	55
7	0.138	77	77	103	103	77
8	0.062	43	43	57	57	43
9	0.064	57	57	76	76	57
10	0.059	34	34	45	45	34
11	0.080	49	49	65	65	49
12	0.088	54	54	71	71	54
13	0.087	65	65	87	87	65
14	0.126	65	65	87	87	65
15	0.072	41	41	55	55	41
16	0.102	54	54	72	72	54
17	0.098	60	60	80	80	60
18	0.068	34	34	45	45	34
19	0.200	103	103	137	137	103
20	0.225	84	84	112	112	84
21	0.067	42	42	56	56	42
22	0.097	39	39	52	52	39
23	0.096	55	55	73	73	55
24	0.111	34	34	45	45	34
25	0.128	60	60	79	79	60
26	0.109	82	82	109	109	82
27	0.087	74	74	98	98	74
28	0.053	24	24	33	33	24
29	0.190	126	126	168	168	126

Table 6. Pond input parameters for each subwatershed.

was used to develop the SWAT2012 model with a revision number 635. A threshold of 0% for land use, 5% for soil, and 0% for slope was used to delineate HRUs resulting in 3402 HRUs. Some past studies reported the relationship between watershed response and HRU delineation approach [18, 19].

Subwatershed	WET_FR	WET_NSA (ha)	WET_NVOL 104 (m^3)	WET_MXSA (ha)	WET_MXVOL 104 (m^3)	WET_VOL 104 (m^3)
1	0.0000	0.00	0.00	0.00	0.00	0.00
2	0.0000	0.00	0.00	0.00	0.00	0.00
3	0.0249	65.97	32.99	219.90	109.95	6.60
4	0.0151	46.43	23.22	154.78	77.39	4.64
5	0.0004	1.38	0.69	4.61	2.30	0.14
6	0.0040	12.62	6.31	42.06	21.03	1.26
7	0.0001	0.15	0.08	0.50	0.25	0.02
8	0.0000	0.00	0.00	0.00	0.00	0.00
9	0.0000	0.00	0.00	0.00	0.00	0.00
10	0.0000	0.00	0.00	0.00	0.00	0.00
11	0.0000	0.00	0.00	0.00	0.00	0.00
12	0.0000	0.00	0.00	0.00	0.00	0.00
13	0.0018	7.18	3.59	23.92	11.96	0.72
14	0.0146	39.90	19.95	133.01	66.51	3.99
15	0.0093	27.84	13.92	92.79	46.39	2.78
16	0.0003	0.96	0.48	3.20	1.60	0.10
17	0.0000	0.00	0.00	0.00	0.00	0.00
18	0.0142	37.57	18.79	125.24	62.62	3.76
19	0.0058	15.53	7.77	51.78	25.89	1.55
20	0.0019	3.77	1.89	12.57	6.28	0.38
21	0.0052	17.23	8.62	57.45	28.72	1.72
22	0.0331	70.06	35.03	233.53	116.76	7.01
23	0.0017	5.04	2.52	16.79	8.40	0.50
24	0.0040	6.33	3.16	21.09	10.54	0.63
25	0.0000	0.00	0.00	0.00	0.00	0.00
26	0.0081	31.88	15.94	106.25	53.13	3.19
27	0.0002	0.81	0.41	2.70	1.35	0.08
28	0.0060	14.39	7.20	47.97	23.99	1.44
29	0.0364	127.13	63.56	423.75	211.88	12.71

Table 7. Wetland input parameters for each subwatershed.

2.4. Calibration and validation

Before calibrating a model, sensitivity analysis is usually performed to reduce the number of parameters. Latin hypercube (LH) one-at-a-time (OAT) method [20] was used to identify the sensitive parameters that might affect the output results. A total of 22 flow parameters were tested, and the following 12 were found sensitive: SOL_AWC, CN2, ALPHA_BF, SOL_K, CH_N2, CH_K2, CANMX, RCHRG_DP, SURLAG, GW_DELAY, OV_N, and GW_REVAP.

The model calibration period was from 1987 to 2006 and the validation period was from 2007 to 2012. The first 3 years of calibration period were selected as a warm-up period so that the model parameters can be initialized. The calibration started with baseflow followed by surface flow adjusting related parameters affecting baseflow and surface flow. The SWAT Check tool [21] was used before calibration to make sure that the simulated outputs were within the reasonable ranges. The Load Estimator (LOADEST) tool [22] was used on a water quality dataset available from Sept 2011 to Dec. 2012 at Hattieville and Apr. 2012 to Dec. 2012 at Morrilton. The regression coefficients were found to be statistically significant ($p < 0.05$) at Hattieville and Morrilton for sediment, total phosphorus, and nitrate nitrogen. The performance of the model was determined mainly using the coefficient of determination (R^2).

3. Results and discussion

3.1. Calibration and validation results

Various SWAT parameters that were calibrated along with their parameter ranges and final calibrated values can be seen in **Table 8**. The annual calibrated R2 for the total, surface, and

File/ parameter	Definition	MIN	MAX	Units	Calibrated value	Notes
.bsn						
ESCO	Soil evaporation compensation factor	0	1		0.95	Based on water balance
EPCO	Plant uptake compensation factor	0	1		1	Based on water balance
.gw						
GW_DELAY	Groundwater delay	0	500		2	Calibrated value
ALPHA_BF	Baseflow alpha factor	0	1	Days	0.0932	Baseflow separation factor
GW_REVAP	Groundwater "revap" coefficient	0.02	0.2		0.072	Calibrated value
REVAPMN	Threshold depth of water in the shallow aquifer for "revap" to occur	0	1000		750	Calibrated value
RCHRG_DP	Deep aquifer percolation fraction	0	1		0.06	Calibrated value
GWQMN	Threshold depth of water in the shallow aquifer required for return flow to occur	0	5000	mm	800	Calibrated value
.rte						
CH_N2	Manning's "n" value for the main channel	−0.01	0.3		0.014	Calibrated value

File/parameter	Definition	MIN	MAX	Units	Calibrated value	Notes
CH_K2	Effective hydraulic conductivity	−0.01	500	mm/hr	6	
.hru						
CANMX-Forest	Maximum canopy storage	0	100	mm	6	Wu et al., [23]
CANMX-Ag	Maximum canopy storage	0	100	mm	2.8	
CANMX-Pasture	Maximum canopy storage	0	100	mm	4	
CANMX-Urban	Maximum canopy storage	0	100	mm	0.1	
SURLAG	Surface runoff lag time	1	24	Days	2	Calibrated value
HRU_SLP	Average slope steepness	0	1	m/m	Reduce by 10%	Based on identified high sediment yield on high-slope agricultural HRUs
.mgt						
CN2	SCS runoff curve number for moisture condition II	35	98		CN + 1	Calibrated value
.sol						
SOL_AWC	Soil available water capacity	0	1	mm/mm	SOL_AWC × 1.13	Calibrated value

Table 8. SWAT model parameter ranges and the final calibrated values.

baseflow was 0.83, 0.85, and 0.16. The validated R2 was 0.91, 0.93, and 0.60 for the total, surface, and baseflow. The monthly calibrated R2 was 0.73, 0.73, and 0.54 and validated R2 was 0.84, 0.78, and 0.76 for the total, surface, and baseflow, respectively. The calibration and validation scatter plots for total flow, surface flow, and baseflow can be seen in **Figure 5**. The validated R^2 for water quality was 0.5–0.7 at Hattieville and 0.7–0.87 at Morrilton. The results are within acceptable limits of other modeling studies relating to limited data availability [24, 25].

4. Conclusions

Modeling studies are gaining popularity due to rapidness of insight generation before actually performing field experiments. The initiative led by the Mississippi River Basin focused on analyzing the water quality benefits from intended best management practices with the help of modeling studies. However, merely simulating best management practices will not be able to provide reliable results unless the model has been set up correctly and robust. This chapter focused on the detailed discussion for setting up the model to a point where the model setup procedure can be replicated. The model was set up with all relevant information, and each data

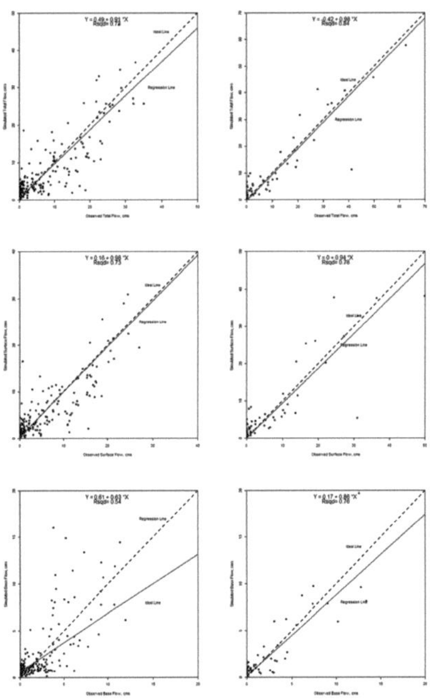

Figure 5. Calibration [left] and validation [right] scatter plots for total flow, surface flow, and baseflow.

preparation step has been explained in detail. The model was calibrated and validated for flow at Hattieville. Due to limited water quality data, the model was validated for sediment, total phosphorus, and nitrate nitrogen at Hattieville and Morrilton. The results were satisfactory and within the ranges reported by previous studies. Results from this study can be used to evaluate the relative effectiveness of MRBI-recommended agricultural BMPs for analyzing pollutant load reductions and improving water quality in similar data-limited watersheds.

Conflict of interest

The authors declare no conflict of interest.

Author details

Gurdeep Singh[1]* and Mansoor Leh[2]

*Address all correspondence to: gurdeep.singh@climate.com

1 The Climate Corporation, St Louis, MO, USA

2 International Water Management Institute, Vientiane, Lao PDR

References

[1] ANRC. The State of Arkansas 2011–2016 Nonpoint Source Pollution Management Plan. Arkansas Natural Resources Commission: Little Rock, Arkansas; 2012

[2] USDA-NRCS. 2010 Mississippi River Basin Healthy Watershed Initiative Projects [Internet]. 2010. Available from: https://www.nrcs.usda.gov/wps/portal/nrcs/detail/ar/programs/landscape/?cid=nrcs142p2_034799 [Accessed: 01-06-2018]

[3] Arnold JG, Srinivasan R, Muttiah RS, Williams JR. Large area hydrologic modeling and assessment part I: Model development. Journal of the American Water Resources Association. 1998;**34**(1):73-89. DOI: 10.1111/j.1752-1688.1998.tb05961.x

[4] Neitsch S, Arnold J, Kiniry J, Williams J. Soil & water assessment tool theoretical documentation version 2009. Texas Water Resources Institute; **2011**:1-647. DOI: 10.1016/j.scitotenv.2015.11.063

[5] Singh G, Saraswat D, Sharpley A. A Sensitivity analysis of impacts of conservation practices on water quality in L'Anguille River Watershed, Arkansas. Water. 2018;**10**(4):443. DOI: 10.3390/w10040443

[6] Singh G, Saraswat D. Development and evaluation of targeted marginal land mapping approach in SWAT model for simulating water quality impacts of selected second generation biofeedstock. Environmental Modelling and Software. 2016;**81**:26-39. DOI: 10.1016/j.envsoft.2015.12.001

[7] Kumar E. SWAT model simulation of bioenergy crop impacts on water quality in cache river watershed [thesis]. Fayetteville, Arkansas: University of Arkansas; 2015

[8] Pai N, Saraswat D, Daniels M. Identifying priority subwatersheds in the Illinois River drainage area in Arkansas watershed using a distributed modeling approach. Transactions of the ASABE. 2011;**54**:2181-2196. DOI: 10.13031/2013.40657

[9] Gassman PW, Reyes MR, Green CH, Arnold JG. The soil and water assessment tool: Historical development, applications, and future research directions. Transactions of the ASABE. 2007;**50**:1211-1250. DOI: 10.13031/2013.23637

[10] Singh G, Kumar E. Input data scale impacts on modeling output results: A review. Journal of Spatial Hydrology. 2017;**13**(1):1-10. Available from: http://spatialhydrology.net/index.php/JOSH/article/view/280

[11] Bracmort KS, Arabi M, Frankenberger JR, Engel BA, Arnold JG. Modeling long-term water quality impact of structural BMPs. Transactions of the ASABE. 2006;**49**:367-374. DOI: 10.13031/2013.20411

[12] Chaubey I, Chiang L, Gitau MW, Mohamed S. Effectiveness of best management practices in improving water quality in a pasture-dominated watershed. Journal of Soil and Water Conservation. 2010;**65**:424-437. DOI: 10.2489/jswc.65.6.424

[13] Benham BL, Baffaut C, Zeckoski RW, Mankin KR, Pachepsky YA, Sadeghi AM, et al. Modeling bacteria fate and transport in watersheds to support Tmdls. Transactions of the ASABE. 2006;**49**:987-1002. DOI: 10.13031/2013.21739

[14] Borah DK, Yagow G, Saleh A, Barnes PL, Rosenthal W, Krug EC, et al. Sediment and nutrient modeling for TMDL development and implementation. Transactions of the ASABE. 2006;**49**:967-986. DOI: 10.13031/2013.21742

[15] Singh G. A watershed scale evaluation of selected second generation biofeedstocks on water quality [MS thesis]. Fayetteville, Arkansas: University of Arkansas; 2012. Available from: https://scholarworks.uark.edu/etd/638/

[16] Singh G. Evaluation of conservation practices through simulation modeling and tool development [PhD diss.]. Fayetteville, Arkansas: University of Arkansas; 2015. Available from: https://scholarworks.uark.edu/etd/1431/

[17] Arnold J, Allen P. Automated methods for estimating baseflow and ground water recharge from streamflow records. Journal of the American Water Resources Association. 1999;**35**(2):411-424. DOI: 10.1111/j.1752-1688.1999.tb03599.x

[18] Cho J, Lowrance RR, Bosch DD, Strickland TC, Her Y, Vellidis G. Effect of watershed subdivision and filter width on swat simulation of a coastal plain watershed1. Journal of the American Water Resources Association. 2010;**46**:586-602. DOI: 10.1111/j.1752-1688.2010.00436.x

[19] Tripathi MP, Raghuwanshi NS, Rao GP. Effect of watershed subdivision on simulation of water balance components. Hydrological Processes. 2006;**20**(5):1137-1156. DOI: 10.1002/hyp.5927

[20] van Griensven A, Meixner T, Grunwald S, Bishop T, Diluzio M, Srinivasan R. A global sensitivity analysis tool for the parameters of multi-variable catchment models. Journal of Hydrology. 2006;**324**:10-23. DOI: 10.1016/j.jhydrol.2005.09.008

[21] White MJ, Harmel RD, Arnold JG, Williams JR. SWAT check: A screening tool to assist users in the identification of potential model application problems. Journal of Environmental Quality. 2014;**43**(1):208-214. Available from: https://dl.sciencesocieties.org/publications/ jeq/abstracts/43/1/208

[22] Runkel RL, Crawford CG, Cohn TA. Load Estimator (LOADEST): A FORTRAN Program for Estimating Constituent Loads in Streams and Rivers. Vol. 4. Tech. Methods. Colorado: U.S. Geological Survey Circular, U.S. Department of the Interior; 2004. p. 69

[23] Wu K, Xu YJ. Evaluation of the Applicability of the SWAT Model for Coastal Watersheds in Southeastern Louisiana. Journal of the American Water Resources Association. 2006;**42**: 1247-1260. DOI: https://doi.org/10.1111/j.1752-1688.2006.tb05298.x

[24] Matamoros D, Guzman E, Bonini J, Vanrolleghem PA. AGNPS and SWAT model calibration for hydrologic modelling of an ecuadorian river basin under data scarcity. In: Ostfeld A, Tyson JM, editors. River Basin Restoration and Management. London, UK: IWA Publishing; 2005. pp. 71-78

[25] Panagopoulos Y, Makropoulos C, Baltas E, Mimikou M. SWAT parameterization for the identification of critical diffuse pollution source areas under data limitations. Ecological Modelling. 2011;**222**:3500-3512. DOI: 10.1016/j.ecolmodel.2011.08.008